BASIC FACTS ON MATHEMATICS

BASIC FACTS ON MATHEMATICS

SUDHIR KUMAR
M.Sc., Ph.D.

ANMOL PUBLICATIONS PVT. LTD.
NEW DELHI-110002 (INDIA)

ANMOL PUBLICATIONS PVT. LTD.
4374/4B, Ansari Road, Daryaganj
New Delhi-110 002

Basic Facts on Mathematics

First Edition 1995

Reprint 1996, 1997, 1998

Reprint 1999

ISBN 81-7488-081-X

PRINTED IN INDIA

Published by J. L. Kumar for Anmol Publications Pvt. Ltd., New Delhi-110 002 and Printed at Mehra Offset Press, Delhi.

Preface

The book **Basic Facts on Mathematics** is especially for the benefit of the students in undergraduate and postgraduate classes. It would be quite useful for all types of readers. It gives the meanings of various words in a simple and comprehensive manner. The words included in the book are those which we come across most frequently and so it is likely to improve the vocabulary and general awareness of the reader. The book has been written in a clear and lucid style. To provide more clarity to some terms diagrams have been provided alongwith them.

The matter has been drawn from different sources for the compilation and the author is highly thankful to all such persons whose works have been consulted while preparing this book.

Comments on the book are most welcome.

Author

A

Abacus. A calculating device consisting of balls strung on wires or rods set in a frame. It is probably of Babylonian origin but its use declined in Europe with the introduction of Arabic numerals in about the 10th century AD. It is still in use in the Middle East and Japan.

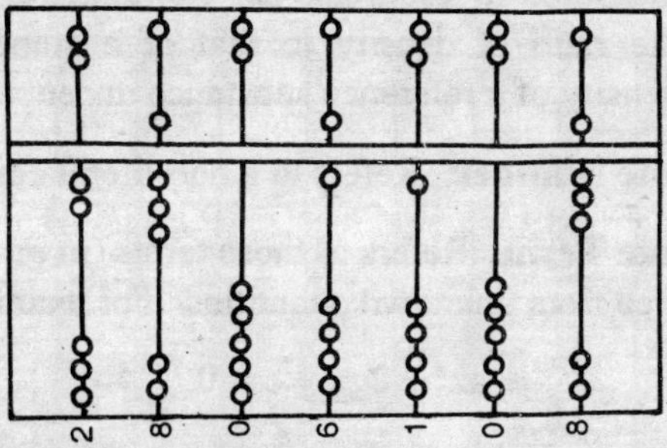

Fig. A-1.

Abel's Integral Equation. A particular type of **Volterra Equation** of the form

$$f(t) = \int_0^t (t-s)^{-\alpha}\, \phi(s)ds \qquad (0 < \alpha < 1).$$

The solution of the equation is

$$\phi(t) = \frac{\sin \pi\alpha}{\alpha} \frac{d}{dt} \int_0^t (t-s)^{\alpha-1}\, f(s)ds.$$

The equation, with $\alpha = 0.5$, has particular application to the time of fall of a particle along a smooth curve in the vertical plane.

Abel's Test (for Convergence). If Σu_n converges and $a_1, a_2, a_3, \ldots$ is a decreasing sequence of positive terms, then $\Sigma a_n u_n$ is convergent.

Abinitio. *From first principles* (by difinition). Generally used in calculus. Used in establishing a result or formula from definition. It is especially used in obtaining derivatives or integrals. For example, differentiate x^n abinitio.

Abscissa. Refers to the horizontal or x-coordinate in a two dimensional rectangular Cartesian coordinate system.

Absolute. Refers to a number or measurement which does not depend on a standard reference value e.g., absolute density is measured in kilograms per cubic metre, but relative density is the ratio of density to that of a standard density (i.e., the density of a reference substance under standard conditions).

Absolute Number. Refers to a non-literal constant.

Absolute Term. Refers to those terms (in an expression) which are free from unknown quantities. For example, in the equation

$$ax^2 + 2gx + c = 0;$$

c is the absolute term.

Absolute Value. *Numerical Value*. Refers to the value of a quantity without any reference to its sign or direction.

Abstract Number. A number which does not refer to any specific quantity. For example, the number 'three' when it does not refer to three objects, quantities, etc., but simply the abstract concept of 'three'.

Abstract Space. A formal mathematical system made up of undefined objects and axioms of a geometric nature e.g., Euclidean spaces, metric spaces, topological spaces, and vector spaces.

Accumulation Points. *Cluster point, Limit point.* This is the limit of a sequence of points in a give set. Thus, in set

$\left\{1, \frac{1}{2}, \frac{1}{3}, \frac{1}{4}, ...\right\}$, the point 0 is an accumulation point.

Accuracy. Correctness, generally referring to numerical computations. The **accuracy of a table** may mean either : (1) The number of significant digits that appear in the numbers in the table (e.g., in the mantissas of a logarithm table); (2) the number of correct places in computations done with the table. (This number of places changes with the form of computations, since errors may repeatedly combine so as to become of any size whatever.)

Accurate. Means exact, precise, without error. An accurate statement one that is correct or true and an accurate computation is one that contains no numerical errors. **Accurate to a certain decimal place** means that all digits preceding and including the given place are correct and the next place has been made zero if less than 5 and 10 if more than 5 (if it is equal to 5, the most usual convention is to call it zero or 10 as is necessary to leave the last digfit even), e.g., 1.26 is accurate to two places if obtained from 1.264 or 1.256 or 1.255.

Action at a Distance. An effect wherein one body affects another through space having no apparent contact or transfer between them.

Actuary. A mathematician employed by an insurance company to calculate the premiums payable on policies. The calculations are based on statistically determined risks and eventualities (e.g., sickness, life expectancy). In the UK qualifications are awarded by the Institute of Actuaries through examination.

Addition Formulae. Equations that express trigonometric functions of the sum or difference of two angles in terms of separate functions of the angles :

$$\sin(x+y) = \sin x \cos y + \cos x \sin y$$

$$\sin(x-y) = \sin x \cos y - \cos x \sin y$$

$$\cos(x+y) = \cos x \cos y - \sin x \sin y$$

$$\cos(x-y) = \cos x \cos y + \sin x \sin y$$

$$\tan(x+y) = (\tan x + \tan y)/(1 - \tan x \tan y)$$

$$\tan(x-y) = (\tan x - \tan y)/(1 + \tan x \tan y)$$

These are used to simplify trigonometric expressions, e.g., in solving an equation. From the addition formulae the following formulae can be obtained :

The double-angle formulae :

$\sin(2x) = 2 \sin x \cos x$

$\cos(2x) = \cos^2 x - \sin^2 x$

$\tan(2x) = 2 \tan x/(1 - \tan^3 x)$

The half-angle formulae:

$\sin(x/2) = \sqrt{[(1 - \text{cis } x)/2]}$

$\cos(x/2) = \sqrt{[(1 + \cos x)/2]}$

$\tan(x/2) = \sin x/(1 + \cos x) = (1 - \cos x)/\sin x$

The product formulae :

$\sin x \cos y = 1/2 [\sin(x + y) + \sin(x - y)]$

$\cos x \sin y = 1/2 [\sin(x + y) - \sin(x - y)]$

$\cos x \sin y = 1/2 [\cos(x + y) + \cos(x - y)]$

$\sin x \sin y = 1/2 [\cos(x - y) - \cos(x + y)]$

Additive, Additive Function. A function of f that has the property that f(x + y) is defined and equals f(x) + f(y) whenever f(x) and f(y) are defined. A *continuous* additive function is always *homogeneous*. A function f is **subadditive** or **superadditive** according as

$$f(x_1 + x_2) \leq f(x_1) + f(x_2),$$

or $$f(x_1 + x_2) \geq f(x_1) + f(x_2),$$

for all x_1, x_2 and $x_1 + x_2$ in the domain of f (this domain is generallly taken to be an interval of the form $0 \leq x \leq a$).

Adiabatic Curves. Refer to curves showing the relation between pressure and volume of substances which are considered to have adiabatic expansion and contraction.

Adinfinitum. Means continuing without end (according to some law); denoted by three dots, as ...; used, mostly in writing infinite series, infinite sequences and infinite products.

Agnesi, Witch of

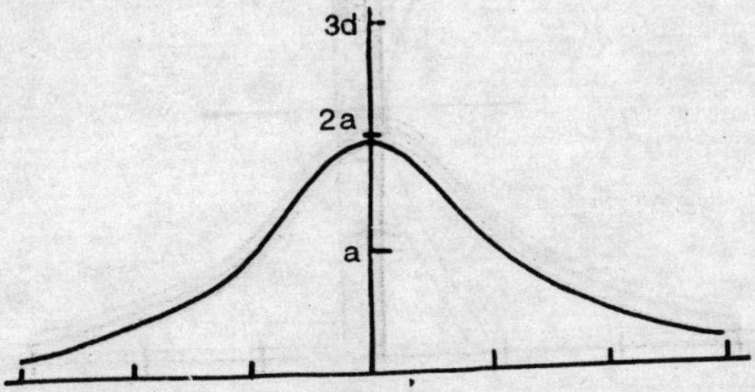

Fig. A-2.

A curve whose equation $x^2y = 4a^2(2a - y)$.

Adjacent Angles. Refers to two angles with a common side and common vertex and lying on opposite sides of their common side. In the figure A-3, AOB and BOC are adjacent angles.

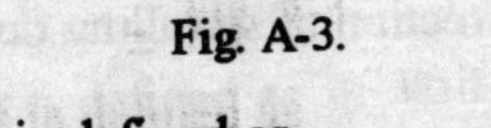

Fig. A-3.

Airy's Integral. This is defined as

$$Ai(z) = \frac{1}{2\pi j}\int_L \exp(tz - t^3/3)\,dt = \frac{1}{\pi}\int_0^\infty \cos(sz + s^3/3)\,ds$$

where L is a contour as shown in the diagram. It is the integral solution of the differential equation

$$\frac{d^2y}{dz^2} - zy = 0.$$

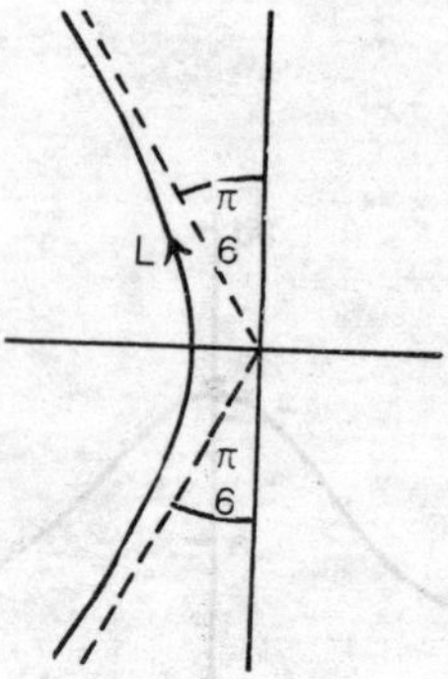

Fig. A- 4.

Airy's Points. The optimum points at which a bar must be suspended horizontally to make bending a minimum. The distance apart of the point is

$$\frac{1}{\sqrt{(n^2 - a)}}$$

where l is the length of the bar and n the number of supports.

Airy's Spirals. The spirals of light which can be produced when convergent polarized light is passed through two plates cut from right-handed and left-handed quartz (enantiomorphic forms of the crystal) and observed with crossed Nicol prisms (i.e., prisms to separate the ordinary and extraordinary rays). The spirals arise from the rotation of the plane of polarization of the light and the direction of spiralling depends on which quartz plate is positioned first.

Aitken's Formula. If a series of numbers u_1, u_2, ..., u_n, u_{n+1}, u_{n+2}, . . . is expected to converge slowly to an unknown limit u, then u can be estimated by assuming

$$u_n = u = e; \; u_{n+1} = u - ke; \; u_{n+2} = u - k^2e \; (k < 1)$$

whence, by eliminating k and e :

$$u = \frac{u_n + 2u_n - u^2_{n+1}}{u_{n+2} - 2u_{n+1} + u_n} = u_{n+2} - \frac{(u_{n+2} - u_{n+1})^2}{u_{n+2} - 2u_{n+1} + u_n}$$

Alexander's Subbase Theorem. It states that a topological sace is *compact* if and only if there is a subbase S for its topology that has the property that, whenever the union of a collection of members of S contains X, the X is contained in the union of a finite number of members of this collection.

Algebra. The branch of mathematics that uses symbols to represent unknown quantities. The first treatise on the subject was written by Diophantus of Alexandria in the 3rd century AD and the name derives from the Arabic al-jar, a term used by the mathematician al-Khwarizmi to denote the addition of equal quantities to both sides of an equation and later adopted as the name for the whole subject. Algebra was used in ancient Babylon, Egypt, and India and brought to Europe by the Arabs. In classical algebra symbols, such as x and y, represent ordinary numbers and the central part of the subject is the study of algebraic equations. Modern, or abstract, algebra is concerned with any system of quantities that obey a particular set of general rules and relationships. Such systems may or may not obey the *commutative laws or even the *associative laws that hold in arithmetic.

Algol (algorithmic language). A computer-programming language. It is used to express mathematical and scientific problems in a way that can be processed by computer. Algol is a high-level language, i.e., statements made in it resemble English rather than a computer notation.

Alternate Angles. A pair of equal angles formed by parallel lines and a third line crossing them $\angle 1$ and $\angle 2$ are alternate angles.

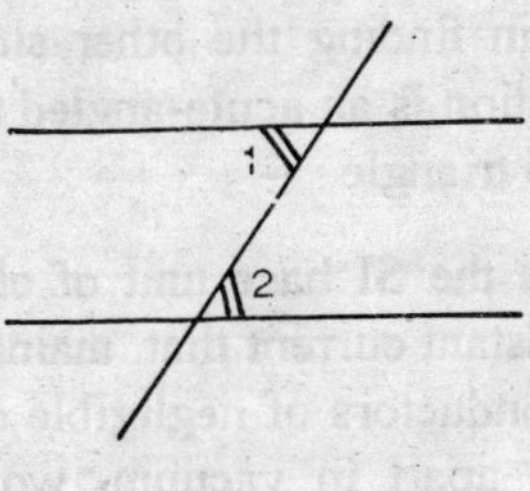

Fig. A-5.

Alternating Series. Refers to a series wherein the terms are alternately positive and negative, e.g.,

$$S_n = -1 + 1/2 - 1/3 + 1/4 - \ldots + (-1)^n/n$$

Such a series is convergent if the absolute value of each term is less than the preceding one. The given series above is a convergent series.

An alternating series can be constructed form the sum of two series, one having positive terms and other having negative terms. In this case, if both are convergent separately then the alternating series is also convergent, even if the absolute value of each term is not always smaller than the one before it. For instance, the series :

$$S_n' = 1/2 + 1/4 + 1/8 + \ldots + 1/2^n$$

and

$$S_n'' = -1/2 - 1/3 - 1/4 - 1/5$$

Altitude. The angular distance of an asrtronomical body above or below an observer's horizon. It reaches a maximum of 90° when the body is directly overhead. It is used with the angular distance **azimuth**, measured eastwards along the horizon from the direction of N, to specify the position of an astronomical body on the *celestial sphere.

Ambiguous. Having more than one possible meaning, value, or solution e.g., if we are given two sides and one angle other than the included angle of a triangle, it may give rise to an ambiguous case in finding the other side and angles of the triangle. One solution is an acute-angled triangle and the other is an obtuse-angle triangle.

Ampere Symbol. It is the SI base unit of electric current and is defined as the constant current that, maintained in two straight parallel infinite conductors of negligible circular cross-section placed one metre apart in vacuum, would produce a force between the conductors of 2×10^{-7} newton per metre.

Analog Computer. A type of computer wherein data is represented by a physical quantity generally voltage that can change continuously. The data taken from some problem may be a changing temperature or pressure in a system. The data in the analog computer gets converted into its voltage analogue. The resulting voltage is proportional to the result of the operation, that is recorded.

Analog computers can perform much more complicated computations that digital computers but are not so accurate and are less flexible in their use.

A-NALY-SIS, [*pl.* analysis]. Refers to that part of mathematics which uses, for the most part, algebraic and calculus methods—as different from such subjects as synthetic geometry, number theory, and group theory.

Analysis of Problem. Refers to the exposition of the principles involved; a listing, in mathematical language, of the data provided in the statement of the problem, other related data, the end sought, and the steps to be taken.

Analytic Geometry. The study of geometrical relations by algebraic methods. Geometrical figures are placed in a *coordinate system, each point in the figure being represented by its coordinates, which satisfy an algebraic equation. Also known as coordinate geometry or Cartesian geometry, after its inventor Rene *Descartes.

Angle (plane angle). Refers to the spatial relationship between two straight lines, i.e., when a straight line rotates about a fixed point on it in a plane the angle is the measure of rotation. According to the convention, an anticlockwise rotation gives a positive angle while a clockwise rotation a negative angle. Angle between two parallel lines is zero. Angles are measured in digrees, or in radians.

1 revolution = 360 degrees

1 degree = 60 minutes (first minute parts)

1 minute = 60 seconds (second minute parts).

A right angle is 90° while a straight line forms two right angles, i.e., 180° or π radians.

Angle at Centre. Refers to the angle between any two radii of a circle. In the figure the arc S is said to subtend an angle θ at the centre. The size of the angle at the centre is directly proportional to the length of the arc.

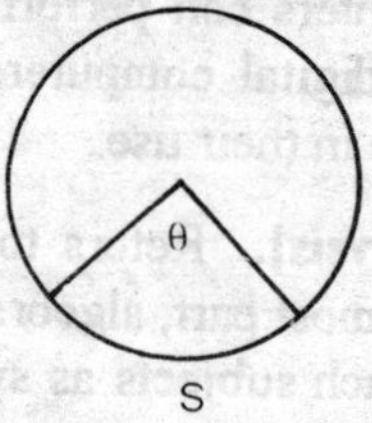

Fig. A-6.

Angle at the Circumference. Refers to the angle between any two chords of a circle, having common point on the circumference of the circle. In the figure arc S is said to subtend an angle β at the circumference. For a given arc the angle at every point of the circumference will be the same. The angle at the circumference is half the angle at the centre for a given arc.

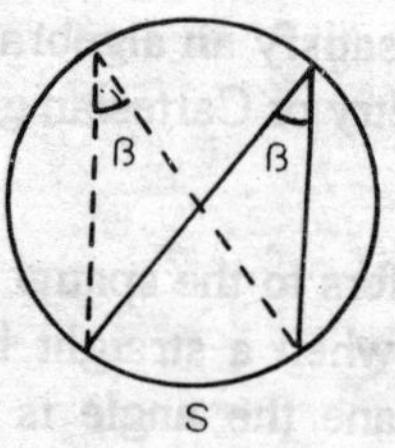

Fig. A-7.

Angle between Two Lines. Angle θ between two lines in a plane with slopes m_1 and m_2, is given by

$$\tan\theta = \frac{m_1 \approx m_2}{1 + m_1 m_2}.$$

Angles Made by a Transversal. Refers to the angles formed by a line (the transversal) that cuts two or more other lines. If the transversal t cuts the lines m and n. The angles a, b, c, d' enclosed between lines are **interior angles**; a', b', c, d outwards are **exterior angles**; a and c', and b and d' are the pairs of **alternate-interior angles**; b' and d, a' and c are the pairs of **alternate exterior**; **angles** a' and a, b' and b, c' and c, d' and d' are the **exterior-interior** or **corresponding angles**.

Angle of a Polygon. Refers to an **interior angle** of a polygon whose vertex is a vertex of the polygon, whose sides contain the sides of the polygon which meet at this vertex, and whose measure is equal to the smallest positive measure which describes, a rotation from one side through the interior of the polygon to the other side. **An exterior angle** is an angle whose vertex is a vertex of the polygon, whose sides contain one side (of the polygon) having an endpoint at this vertex and the other side.

Angular Acceleration. Symbol : α Refers to the rotational acceleration of a body about an axis; i.e., the rate of change of angular velocity with time.

$$\alpha = d\,\omega/dt$$

or $$\alpha = d^2\theta/ dt^2$$

where ω is *angular velocity* and θ is angular displacement.

Angular Displacement. Symbol : θ. Refers to the rotational displacement of an object about an axis. If the object (or a point on it moves from point P_1 to point P_2 in a plane perpendicular to the axis, θ is the angle P_1OP_2, where O is the point where the perpendicular plane meets the axis.

Angular Frequency (pulstance). Symbol : ω. The number of complete rotations per unit time., Angular frequency is generally used to describe vibrations. A simple harmonic motion of frequency f can be represented by a point moving in a circular path at constant speed. The foot of a perpendicular from the point to a diameter of the circle moves with simple harmonic motion. The angular Frequency of motion is equal to $2\pi f$, where f is the frequency of the simple harmonic motion.

Angular Levelling. Means determining differences in elevation of points on earth's surface. In the figure OB is horizontal and AB is vertical. Angle α_2 measures the *angle of elevation* of A from O. Angle α_2 measures the angle of depression of O from A.

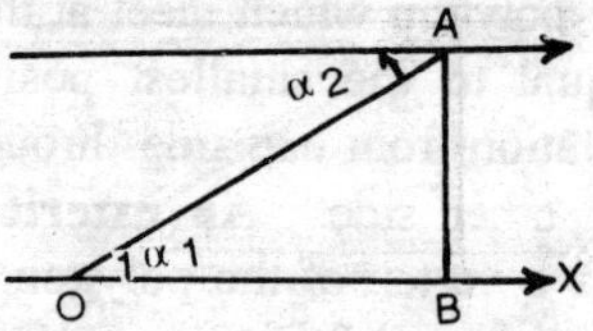

Fig. A-8.

Anharmonic Oscillator. Refers to a system whose vibration, though periodic, cannot be described in terms of simple harmonic motions (i.e., sinusoidal motions). In such cases, the period of oscillation is not independent of the amplitude.

AN-NIHI-LATOR. Refers to the annihilator of a set S is the class of all functions of certain type which *annihilate* S in the sense of being zero at each point of S e.g., if the functions are continuous linear functionals and S is a subset of a formed linear space N, then the annihilator S is the linear subset S' of the first conjugate space N that is made up of all continuous linear functionals which are zero at each point of S.

Annuity. Refers to a pension wherein an insurance company pays the annuitant fixed amount of money in return for sums of money paid to it either in instalments or as a lump sum. An *annuity certain* is paid for a fixed number of years whereas annuity that is payable only while the annuitant is alive.

Annulus. Refers to the region between two concentric circles. The area of an annulus is $\pi (R^2 - r^2)$, where R is the radius of the larger circle and *r* is the radius of the smaller circle.

Antecedent. In logic, refers to the first part of a conditional statement; a proposition or statement which is said to imply another. For example, in the statement "if it is hot then the streets are dry," it is hot is the antecedent.

Applied Mathematics. Refers to the study of the mathematical techniques which are used to solve problems. Strictly speaking it is the application of mathematics to any 'real' system. For example, pure geometry is the study of entities—lines, points, angles, etc., which is based on certain axioms. The use of Euclidean geometry in surveying, architecture, navigation, or science is applied geometry. The term 'applied mathematics' is used particularly for mechanics—the study of forces and motion.

Apollonius' Circle. The locus of the vertex, A, of a triangle of given base BC such that the sides AB and AC are in a given ratio $\lambda : 1$ is a circle with, as diameter, the line joining the points which divide the base BC internally and externally in the ratio $\lambda : 1$.

Apponias Theorem. This theorem relates the length of a median in a triangle to the lengths of tis sides. If AD is the median of the triangle ABC, of length m, then

$$b^2 + c^2 = 2m^2 + 2\,(a/2)^2.$$

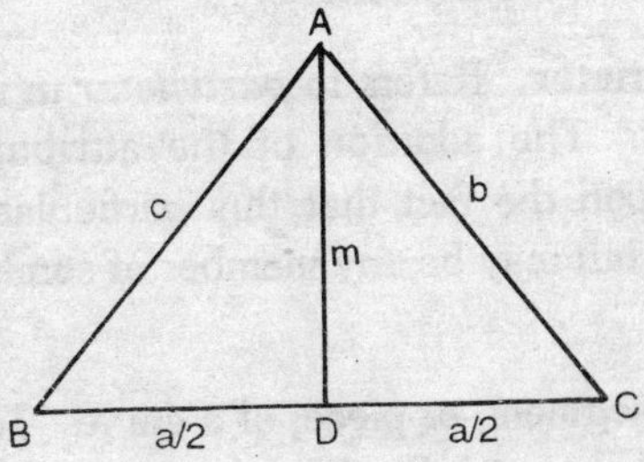

Fig. A-9.

Approximation. Refers to a number that is taken as a close estimate of another number, or the method of finding such an approximate number. The symbol used for approximation is $\approx$ or $\cong$. For example, $\pi \cong 3.1416$, $\sqrt{2} \approx 1.4142$.

A Priori Knowledge. Refers to knowledge got from pure reasoning from cause to effect, as opposed to empirical knowledge (knowledge obtained from experience); Knowledge that has its origin in the mind and is considered to be) quite independent of experience.

A Priori Reasoning. Refers to reasoning which arrives at conclusions from definitions and assumed axioms or principles.

Arabic Numerals. The number symbols 0, 1, 2, 3, 4, 5, 6, 7, 8, 9. They are believed to have originated in India and were introduced into Europe by Arabs in about the 10th century AD.

Arbitrary Assumption. Refers to an assumption constructed at the pleasure of the individual without any regard to its being consistent either with the laws of nature or (sometimes) with accepted mathematical principles.

Arbitrary Function in the Solution of Partial Differential Equations. Refers to a symbol that stands for an unspecified function in expression which satisfies the differential equation whatever function (or some specified type) may be substituted for the symbol e.g. $z = xf(y)$ is a solution of $x\,(\partial z/\partial x) — z = 0$, if f is any differentiable function.

Arbitrary Parameter. Refers to *parameter* in its most commonly used sense. The addition of the attribute *arbitrary* laws emphasis upon the fact that this particular parameter is not determined, but may be any member of some set (e.g., any real number).

Arc. Refers to a segment, or piece, of a curve. *Tech* (1) The image of a closed interval [a, b]under a one-to-one continuous transformation; i.e., a simple curve which is not closed. (2) A curve which is not a closed curve. If a curve is the continuous image of the interval [a, b] then an arc of the *curve* is any arc which is the image of an interval [c, d] contained in [a, b].

Arc Cosine. Refers to the arc cosine of a number x, is an angle (or number) whose cosine is x, written $\cos^{-1} x$ or arc cos x. e.g., arc cos 1/2 is 60°, 300°, or in general n 360° ± 60°.

Archimedean Polyhedra. These are semi-regular polyhedra, also called Archimedean solids, for which every face is a regular polygon, although the faces are not all of the same kind. The faces must be arranged in the same order around each vertex. There are 13 Archimedean solids although two of these 13 can exist in enantiomorphic forms : i.e., in right-handed and left-handed forms. They can all be described in a sphere. Additionally there are 13 Archimedean duals in which all the faces are congruent and their polyhedral angles regular but not identical. Such solids are referred to as vertically-regular and has an inscribed sphere. Each Archimedean dual can be obtained from a corresponding Archimedean solid by replacing every vertex of the latter by the tangent plane to the sphere in which it can be described.

Area. Symbol : A. The extent of a plane figure of a surface, measured in units of length squared. The SI unit of area is the square metre (m^2). The area of a rectangle is given by the product of its length and breadth. A more complex-shaped area can be formed by superimposing a square grid and counting the number of whole and part squares that it covers.

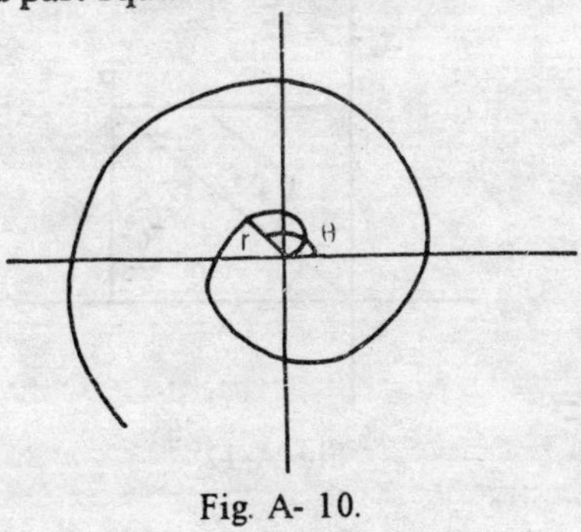

Fig. A- 10.

Area Under the Curve. Refers to the area bounded by the curve y = f (x), two ordinates x = a and x = b, and the x-axis. It is obtained by taking help of calculus.

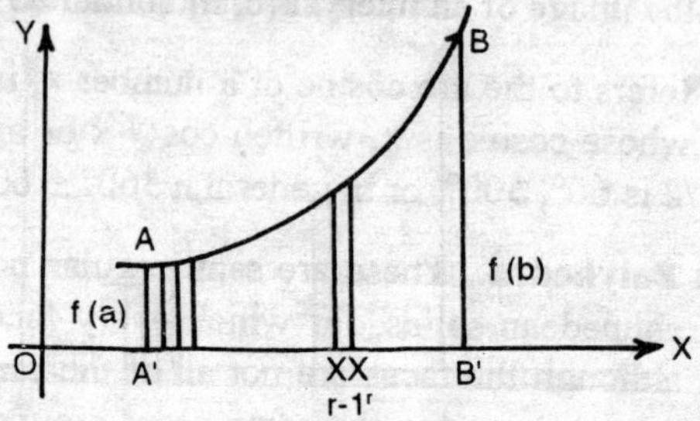

Fig. A-11.

The interval (a, b) is divided into n parts so that rth sub-interval is $\delta x_r = x_r - x_r - x_{n-1}$. The area under the curve is approximated by

$\sum_{r=1}^{n} y_r \delta x_r \; (= \Sigma y \delta x)$ and is actually equal to $\lim_{n \to \infty} \sum_{r=1}^{n} y_r \delta x_r$

represented by $\int_a^b y\,dx \left(= \int_a^b f(x)\,dx \right)$.

Argand Diagram. The representation of a complex quantity.

$$z = x + iy$$

where x is the real and y the imaginary part, by a joint P whose Cartesian coordinates are x, y. If z is expressed in De Moivre's form.

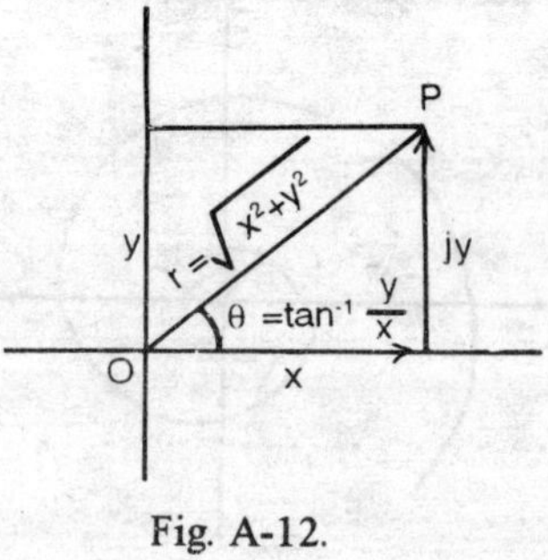

Fig. A-12.

$$z = re^{j\theta} = r(\cos\theta + j\sin\theta)$$

$r = \sqrt{(x^2 + y^2)}$, the length of the line from the origin O to P, and $\theta = \tan^{-1}(y/x)$ where θ is the angle between OP and the x-axis.

Argument (amplitude). In a complex number written in the form $r(\cos\theta + i\sin\theta)$, the angle θ is the argument. It is thus the angle that the vector representing the complex number forms with the horizontal axis in an Argand diagram.

Arithmetic. The branch of mathematics that deals with elementary theories of numbers, measurement, and computation. The fundamental operations of arithmetic are addition, subtraction, multiplication and division. Addition and multiplication are assumed to obey the * associative law, the * commutative law, and the *distributive law. Other operations in arithmetic include extracting roots, raising a number to a power, and taking *logarithms. Arithmetic is also concerned with fractions and the various number systems, such as the *decimal system and the *binary system.

Arithmetic Mean(s). Refers to the terms of an arithmetic progression, that lie between the first and the last and are called arithmetic means of the first and the last terms. To insert n means between two numbers a and b means having n + 2 numbers in arithmetic progression having common difference equal to $\frac{b-a}{n+1}$. Thus, the n arithmetic means in between a and b are $a + \frac{b-a}{n+1}$, $a + 2\left(\frac{b-a}{n+1}\right)$......., $a + n\left(\frac{b-a}{n+1}\right)$ The **arithmetic mean** of a and b is given by $\frac{a+b}{2}$.

Arithmetic Progression (*or* arithmetic sequence). A sequence of numbers in which successive terms have a constant difference, for example, 1, 4, 7, 10, 13........

Array. Refers to numbers or other items of information, like those in a list or table since computing, an array has its own name, *identifier,* and each member of the array is identified by a subscript used with the identified. An array can be examined by a programme and a particular item of information extracted by using this identifier and subscript.

Artificial Intelligence (AI). The design of computer programs to perform tasks that involve intelligence when carried out by human beings. Tasks performed by AI include playing games, learning, understanding natural language and speech, formulating plans, proving theorems, and visual perception. Perceptual tasks involve unconscious computation in humans, which is hard to simulate. AI programs sometimes simulate human behaviour by building computer models of cognitive processes, but in most cases they adopt any technique that will do the task set or do it better than it has been done before.

Ascoli's Theorem. Let A be an infinite set of functions all of whose domains are the same closed bounded set D in a finite-dimensional Euclidean space (e.g., a closed bounded interval) and whose ranges are set of real numbers. If these functions are *equicontinuous* and there is a number M such that $|f(x)| \leq M$ for all f in A all x in D, then there is a sequence $(_n)$ of distinct members of A that converges uniformly to a continuous function. The following stronger theorem also it true. Let A be a set of functions whose domains are the same separable metric space X and whose ranges are in a metric space Y. If these functions are equicontinuous and for each x (in a dense subset of X the set(f(x) : f — A) is compact, then there is a sequence (f_n) of distinct members of A which converges pointwise to a continuous function f land the convergence is uniform on each compact subset of X.

Assembler. A computer *program that makes up part of the *software of a computer. It converts instructions in a programming language into a form that the machine can follow directly.

Assessed Value. Refers to a value set upon property for the purpose of taxation.

Associate. In a commutative semigroup or ring, associates are members a and b for which there exist members x and y such that a = bx and b = ay.

Associated Radius of Convergence. If the power series

$$\Sigma\, a_{k1, k2, \ldots, kn} z_1^{k_1} z_2^{k_2} \ldots z_n^{kn}$$

converges for $|z_j| < r_i$, $j = 1, 2, \ldots, n$, and diverges for $|z_j| > r_i$, $j = 1, 2, \ldots, n$, where r_j is positive, then the set $r_1, r_2, \ldots, r_n$ is known as a set of associated radii of convergence for the series. e.g., for the series.

$$1 + z_1 z_2 + z_1^2 z_2^2 + \ldots \equiv \frac{1}{1 - z_1 z_2}$$

associated radii are any positive numbers r_1, r_2 with $r_1 r_2 = 1$.

Associative. Refers to method of combining objects two at a time. It is *associative* if the result of the combination of three objects (order being preserved) does not depend on the way in which the objects are grouped. If the operation is denoted by o and the result of combining x and y by x o y, then

$$(x\ 0\ y)0\ z = x\ 0\ (y\ 0\ z)$$

The law can be extended to say that in any sum of several terms any method of grouping may be used (i.e., at any stage of the addition one may add two adjacent tetrms). The associative law for ***multiplication*** states that

$$(ab)\, c = a\, (bc)$$

for any numbers a, b, c. This law can be stated that in any product of several factors any method of grouping may be used (i.e., any stage of the multiplication one may multiply any two adjacent factors).

Associative Law. The mathematical rule, obeyed by addition and multiplication in ordinary arithmetic, that the order in which successive identical operations are performed does not affect the result: for addition a + (b + c) = (a + b) + c; for multiplication a (bc) = (ab)c.

Astronomical Unit (au, AU). Refers to the mean distance between the Sun and the Earth, used as a unit of distance in astronomy for measurements within the solar system. It is about 1.496×10^{11} metres.

Asymptote. Refers to a straight line towards which a curve approaches but never meets. A hyperbola, for example, has two asymptotes. In two-diamensional Cartesian coordinates, the curve having the equation $y = 1/x$ has the lines $x = 0$ and $y = 0$ as asymptotes, since y becomes infinitely small, but never reaches zero, as x increases, and vice-versa

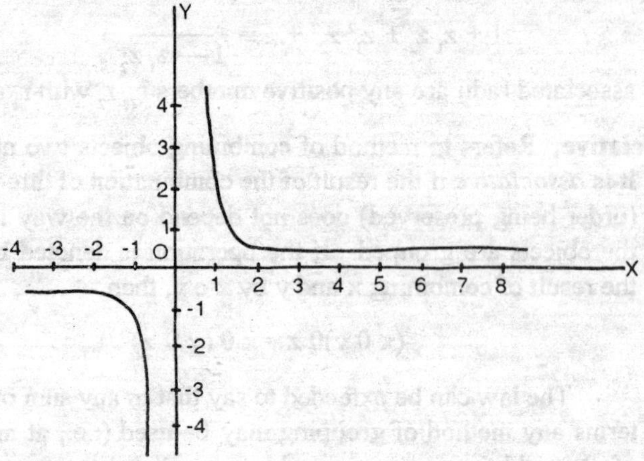

Fig. A-13.

The x-axis and the y-axis are asymptotes to this curve.

Atmospheric Pressure. Refers to the pressure at a point owing to the weight of air above that point. It is equal to the weight of a column of water about 30 ft. high or mercury 760.1 mm high. Its value is about 100 k Pa (100000 newtons per square metres).The atmospheric pressure decreases with elevation above sea-level and with increasing humidity of the atmosphere.

AU (au). See astronomical unit.

Automation. The use of electronic devices controlled by a computer in mechanical processes that would otherwise be controlled by human operators. It has made considerable impact on production engineering in such fields as steel and chemical manufacture. Telecommunications, transport, and mining also employ automated systems.

Average. A representative or middle value of a set of quantities. The arithmetical average (or arithmetical mean) is found by adding the quantities in a set and dividing the total by the number of quantities: the arithmetical average of 7, 8, 13 and 20 is 12 (48 divided by 4). The geometric average (or geometric mean) is found by multiplying together the numbers in a set and extracting a root equal to the number of quantities; the geometric average of 2, 9 and 12 is 6 ($\sqrt[3]{216}$).

Axial Plane. Refers to a fixed referenced plane in three-dimensional coordinate system. For example, in rectangular Cartesian coordinates, the planes defined by $x = 0$, $y = 0$ and $z = 0$ are axial planes. The x-coordinate of a point is its perpendicular distance from the plane $x = 0$, and the y and z coordinates are the perpendicular distances from the $y = 0$ and $z = 0$ planes respectively.

Axiom. An assumption or principle, used to prove a theorem, that is itself accepted as true without proof. Some mathematicians reserve the term for an assumption in logic, using postulate for an assumption made in other fields.

Axis. Refers to (1) A line of symmetry in a given figure.

(2) A fixed reference line used in a coordinate system.

(3) A line about which a body or a curve can be considered to rotate or revolve. For example, the centre line of a right circular cone.

B

Barn. Symbol : *b*. Refers to unit of area equal to 10^{-28} square metre. The barn is sometimes used to express the effective cross-sections of atoms or nuclei in the scattering or absorption of particles.

Barrel. Refers to a US unit of capacity used measure solids. It is equal to 7056 cubic inches (0.1156 m^3).

Base. Refers to a side (or a face) of a geometric figure upon which an altitude is constructed or is considered as being constructed. For an expression such as a^n, the quantity a is called the base and n the exponent.

Basic. A computer language developed in the USA in the 1960s. Designed principally for use by inexperienced programmers, it is one of the easiest languages to learn and a widely used, in more recent versions, in personal computers.

Bateman Equations. Equations applicable to a chain of radioactive nuclides denoted subscripts 1 to n. The nth nuclide decays into the (n + 1)th nuclide with a disintegration constant λ_n, and at any time t there are N_n atoms of the nth nuclide. Then

$$\frac{dN_1}{dt} = -\lambda_1 N_1$$

and $$\frac{dN_n}{dt} = \lambda_{n-1} N_{n-1} - \lambda_n N_n.$$

If at t = 0 only the parent substance is present, and $N_1 = N_{-1}^0$, $N_2^0 = N_3^0 = \ldots = N_n^0 = 0$, then the number of atoms of nth member of the chain at time t is

$$N_n(t) = \sum_1^n \frac{\lambda_1 \lambda_2 \ldots \lambda_{n-1} N_1^0 e^{-\lambda nt}}{(\lambda_1 - \lambda_n)(\lambda_2 - \lambda_n) \ldots (\lambda_{n-1} - \lambda_n)}$$

Bayes' Theorem. If B is an arbitrary event whose probability (E) is not zero, and if $A_1, A_2, \ldots A_n$ are a series of mutually exclusive and exhaustive events, then the probability $P(A_j/B)$ of event A_j happening, once B has happened, is given by

$$P(A_j|B) = \frac{P(A_j)\,P(B \mid A_j)}{\sum_{i=1}^{n} P(B \mid A_i)\,P(A_i)}$$

Thus the conditional probability of A_j happening, given B, is proportional to the unconditional probability of A_j multiplied by the conditional probability B, given A_j.

Benefits of an Insurance Policy. The sum or sums that the company promises to pay provided as specified event takes place like the death of the insured or his attainment of a certain age.

Ber Function. The **ber, bei, her, hei, ker** and **kei** functions are defined by the relation :

$$\mathrm{ber}_n(z) \pm i\,\mathrm{bei}_n(z) = J_n(ze^{\pm 3\pi i/4}),$$

$$\mathrm{her}_n(z) + i\,\mathrm{hei}_n(z) = H_n^{(1)}(ze^{3\pi i/4}),$$

$$\mathrm{her}_n(z) - i\,\mathrm{hei}_n(z) = H_n^{(2)}(ze^{-3\pi i/4}),$$

$$\mathrm{ker}_n(z) \pm i\,\mathrm{kei}_n(z) = i \pm {}^{n}K_n(ze^{\pm 3\pi i/4}),$$

where J_n is a *Bessel function,* $H_n(^1)$ and $H_n(^2)$ are *Hankel functions,* and K_n is a *modified Bessel function* of the second kind. The following conventions are followed $\mathrm{ber}_0(z) = \mathrm{ber}(z)$, $\mathrm{bei}_0(z) = \mathrm{bei}(z)$, etc. It follows that :

$$2\,\mathrm{ker}_n(z) = -\pi\,\mathrm{hei}_n(z);$$

$$2\,\mathrm{kei}_n(z) = \pi\,\mathrm{her}_n(z).$$

These six functions are real when n is real and z is real and positive. In particular,

$$\mathrm{ber}\,x = 1 - \frac{x^2}{2^2 \cdot 4^2} + \frac{x^8}{2^2 \cdot 4^2 \cdot 6^2 \cdot 8^2} - \ldots,$$

$$\text{bei } x = \frac{x^2}{2^2} - \frac{x^5}{2^2 . 4^2 . 6^2} + \frac{x^{10}}{2^2 . 4^2 . 6^2 . 8^2 . 10^2} - \ldots,$$

Also $\int_0^x t \text{ ber}(t)\,dt = x \text{ bei}(x)$, $\int_0^x t \text{ bei}(t)\,dt = - x \text{ ber}(x)$, these still being valid if ber is replaced by *ker* and *bei* by *kei*.

Bernoulli. A family of notable Swiss mathematicians and physicists, who, as Flemish Protestants, were driven out of the Netherlands in the 1580s. The most famous was **Daniel Bernoulli** (1700-82), who made important contributions to fluid dynamics, especially his discovery in 1738 of *Bernoulli's principle concerning the speed and pressure of fluid flow. He also attempted to explain the properties of gases at varying temperatures and pressures by regarding the gas as consisting of many tiny particles. His father **Jean Bernoulli** (1667-1748) and uncle **Jacques Bernoulli** (1654-1705) were both eminent mathematicians, who often worked together and jointly developed the calculus of variations. Jean also made discoveries in *probability theory and Jacques in *complex numbers. Daniel's brother **Nicolas Bernoulli** (1695-1726) also contributed to the theory of probability.

Bernoulli Polynomials. Bernoulli polynomials $B_r^{(n)}(x)$ of order n are given by the identity

$$\frac{t^n e^{xt}}{(e^t - 1)^n} = \sum_{r=0}^{\infty} \frac{t^r}{r!} B_r^{(n)}(x)$$

where n = 0, 1, 2,.... In addition, the polynomials $B_r^{(n)}(0) = B_r^n$ are called Bernoulli numbers of order n, except for n = 1 when they are usually simply called **Bernoulli numbers,** B_r.

Bernoulli's Differential Equation. Bernoulli's differential equation is of the form

$$\frac{dy}{dx} + f(x)y = g(x)y^n$$

This equation may be made linear by the substitution

$$y = u^{1/(1-n)}$$

Bernoulli's Equation (of Liquid Motion). Liquid escapes through an orifice of cross-sectional area a with a velocity v given by

$$v^2 = 2gh \frac{A^2}{A^2 - a^2}$$

where A is the surface area, h is the head of liquid and g is the acceleration due to gravity. For large values of A this equation reduces to **Torricelli's Law of Efflux.**

Bernoulli's Theorem (for Probability). A sequence of independent trials, in which the probability of 'success' is the same for each, is termed a series of **Bernoulli trials.** If s is the number of 'successes' in a series of n independent trials and p is the probability of an individual 'success', then Bernoulli's theorem states that the probability P of the relative frequency of 'successes'(= s/n) differing from p by more than ε tends to zero as $n \to \infty$ for any $\varepsilon > 0$; i.e.

$$\lim_{n \to \infty} P \left\{ \left| \frac{s}{n} - p \right| > \varepsilon \right\} = 0 \; (\varepsilon > 0)$$

Bernoulli Trial. Refers to an experiment having two mutually exclusive outcomes.

Bertillon. (1851-1922) was a statistician. As head of the Paris bureau of social statistics, he developed new systems of data analysis. He was particularly interested in the high incidence of alcoholism in France and the decline of the French population in comparison with rates in other countries.

Bertrand's Test. (for Convergence)

If $f(x) = O(x^{\lambda-1})$ or if $f(x) = O(x^{-1} [\ln x]^{\lambda-1})$

then $\int_a^\infty f(x)\,dx$ converges when $\lambda < 0$.

Bessel-Clifford Differential Equation

$$u \frac{d^2y}{du^2} + (n+1) \frac{dy}{du} + y = 0$$

reduces to Bessel's equation (see **Bessel Functions**) with the substitution

$$y = zu^{-n/2} \text{ and } u = x^2/4$$

The equation has a solution of the form

$$y = u^{-n/2} J_n(2\sqrt{u}) + u^{-n/2} N_n(2\sqrt{u}).$$

Bessel Functions

Bessel's equation

$$\frac{d^2z}{dx^2} + \frac{1}{x}\frac{dz}{dx} + \left(1 - \frac{n^2}{x^2}\right) z = 0$$

has as its general solution

$$z = AJ_n(x) + BN_n(x)$$

where $J_n(x)$ and $N_n(x)$ are Bessel functions of order n^n . $J_n(x)$ are Bessel functions of the first kind:

$$J_n(x) = \frac{x^n}{2^n\Gamma(n+1)}\left[1 - \frac{x^2}{2(n+1)} + \frac{x^4}{2^4 2!(n+1)(n+2)} - \frac{x^6}{2^6 3!(n+1)(n+2)(n+3)} + \ldots\right]$$

$N_n(x)$ are Bessel functions of the second kind :

$$N_n(x) = \frac{2}{\pi} J_n(x)\left\{\ln\frac{x}{2} + y\right\} - \frac{1}{\pi}\left(\frac{x}{2}\right)^{-n} \sum_{k=0}^{k=n-1} \frac{(n-k-1)}{k!}\left(\frac{x}{2}\right)^{2k}$$

$$-\frac{1}{\pi}\left(\frac{x}{2}\right)^{n} \sum_{k=0}^{k=x} \frac{(-1)^k}{(n+k)!k!}\left(1 + \frac{1}{2} + \frac{1}{3} + \ldots \frac{1}{k} + 1 + \frac{1}{2} + \frac{1}{3} + \ldots \frac{1}{n+k}\right)\left(\frac{x}{2}\right)$$

γ is **Euler's Constant**. $N_n(x)$ are sometimes referred to as **Neumann's** or **Weber's Bessel Functions of the second kind.**

Bessel Inequality

$$|A| + |B| \geq |A + B|$$

where A and B are two vectors.

Bhaskhara II (1114-c. 1185). Indian mathematician, who was the first to use the decimal system in a written work, invented the + and — convention, and used letters to represent unknown quantities as in modern algebra.

Bhavachakra (*Sanskrit:* wheel of becoming). In Buddhism, an image of the cyclical nature of earthly existence, in the form of a wheel held by the demoin of impermanence. Its segments represent the six possible states into which beings are reborn: the realms of gods, titans, hungry ghosts, humans, animals and demons. At the centre, turning the wheel, are greed, hatred, and delusion, depicted as a cockerel, snake, and pig, biting each other's tails. Around the rim, the 12 stages in the cycle of life are symbolically expressed.

Bienayme-Chebyshev Inequality. If $x_1, x_2, \ldots x_n$ are uncorrelated random variables which have a mean value $\bar{x} \left(= \frac{1}{n} \sum_{i=1}^{n} x_i \right)$ and which are from a distribution having a mean value m and standard deviation σ, then for any $\lambda > 0$.

$$P\left\{ (\bar{x} - m) \geq \frac{\lambda\sigma}{\sqrt{n}} \right\} \leq \frac{1}{\lambda^2}$$

where P represents probability.

Bijection. Refers to a bijection from a set A to set B which is a one to one correspondence between A and B, i.e., a function form A into B that is both an injection and surjection.

Bilinear. A mathematical expression that is linear with respect to each of the two variables concerned. A function f(xy) is bilinear if

$$f(x_1 + x_2, y) = f(x_1, y) + f(x_2, y)$$

and $$f(x, y_1 + y_2) = f(x, y_1) + f(x, y_2).$$

Binary. Means based on two. A binary number is made up of only two different digits, 0 and 1, instead of ten in the decimal system. Each digit repreesents a unit, twos, fours, eights, sixteens, etc., instead of units, tens, hundreds, etc. For example, 2 is written as 10, 3 is 11, 16 ($= 2^4$) is 10 000. Computers calculate using binary numbers. The digits 1 and 0 correspond to on/off or high/low conditions in an electronic switching circuit.

Binary System. A number system that uses only two digits 0 and 1. Numbers are expressed in powers of 2 instead of powers of 10, as in the decimal system. In binary notation, 2 is written as 10, 3 as 11, 4 as 100, 5 as 101, and so on. Computers calculate in binary notation, the two digits corresponding to two switching positions (e.g., on or off) in the individual electronic devices in the logic circuits.

Binomial Coefficient. Refers to the factor multiplying the variable in a term of a binomial expansion. For example, in $(x + y)^2 = x^2 + 2xy + y^2$ the binomial coefficients are 1, 2 and 1 respectively. In general, the rth binomial coefficient in the expansion of $(x + y)^n$ is

$$n!/[r^n (n - r)!].$$

This is written using the notation ^nC_r.

Binomial Expansion. Refers to a rule for the expansion of an expression of the form $(x + y)^n$. x and y can be any real numbers and n is an integer. The general formula, sometimes known as the *binomial theorem,* is

$$(x + y)^n = x^n + nx^{n-1} y + [\{n (n - 1) /2!] x^{n-2} y^2 + .. + y^n$$

when, for example, $n = 2$, then $(x + y)^n = x^2 + 2xy + y^2$. If $n = 3$, then $(x + y)^n = x^3 + 3x^2y + 3xy^2 + y^3$. If y is smaller than x and is less than one, then when n is very large, the first few terms of the expansion are approximately equal to the complete series. For example,

$$(2 + 0.02)^8 = 2^8 + 8 \times 2^7 \times 0.02 + [(8 \times 7) / (2 \times 1)]\, 2^6 \times (0.02)^2 + ... = 256 + 20.48 + 0.7168 + ..$$

or approximately 277.

Binomial Nomenclature. A system devised by Linnaeus in the 18th century for the scientific naming of plants and animals, each species being identified by two internationally recognized Latin names—the name of the genus (written with an initial capital letter) followed by the name of the species. The names are usually written in italics and the specific name may be followed by the author's name, usually abblreviated. Thus the wolf is *Canis lupus* L (for Linnaeus).

Binomial Theorem. The theorem, discovered by Newton in 1676 that the quantity $(a + n)^n$ where n is an integer, can be expanded in a series: $(a + b)^n = a^n + na^{n-1} b + [n (n - 1) a^{n-2} b^2] / 2! + [n (n - 1) (n - 2) a^{n-3} b^3] / 3! + ... + b^n$ where, for example, 3! (called factorial three) is $3 \times 2 \times 1$.

Bistable Circuit. Refers to an electronic circuit which has two stable states. The circuit will remain in one state until the application of a suitable pulse, that will cause it to take up the other states. Bistable circuits are used extensively in computer equipment for strong data and for counting. They generally have two input terminals to which pulses can be applied. A pulse on one input makes the circuit to charge state; it will remain in that state until a pulse on the other input causes it to assume the alternative state. These circuits are called *flip-flops*.

Bit. A binary digit. The basic unit of information in information theory and computer memory stores. It is the amount of information needed to specify one of two alternatives, i.e., to distinguish between 1 and 0 in the *binary notation.

Bolzano-Weierstrass Theorem. A bounded sequence of numbers always contains a convergent sequence. (A sequence is bounded if there is a number A such that, foir all values of n. $| x_n | A$). For instance, consider the bounded but non-convergent sequence $\{x_n\}$ where $x_n = (—1)^n$. This has a sub-sequence $\{x_{2n}\}$ which converges to 1.

Bolza, Problem of. The general problem of finding the class of curves $y_i = y_i (x)$ such that

$$\int_{a(z)}^{b(z)} F_i x, y, y^j) \, dx + G_i(z)$$

are minima subject to constraints of the form

$$\Phi_j(x, y, y') = 0.$$

For fixed points, $G_i (z) = 0$, and the problem becomes an isoperimetric one; for example, the problem of finding the path between two points of a particle of pre-assigned initial velocity under constant acceleration due to gravity.

Boolean Algebra. An algebra of sets used to enunciate logical propositions and their consequences. The algebra has two binary operations called addition and multiplication. The algebra may be used to represent binary logic and can be applied to computer logic, switching networks and so on.

Bound 1. In a set of numbers, refers to a value beyond which there are no members of the set. A *lower bound* is less than or equal to every number in the set. *An upper bound* is greater than or equal to every number in the set. The *least upper bound* is the lowest number that is greater than or equal to the set and the *greatest lower bound* is the highest number that is greater than or equal to the set. For example, the set (0.6, 0.66, 0.666,......) has a least upper bound of 2/3.

2. A bound of a function is a bound of the set of values that the function can take for the range of values of the variable. For example, if the variable x can be any real number, then the function $f(x) = x^2$ has a lower bound of 0.

Bound of a Function. Bound of the set of all possible values of the function of a variable. For example, the trigonometric function sine and cosine are bounded by +1 (the l.u.b.) and —1 (g.l.b.). The exponential function 3^x is bounded by 0 (g.l.b.) as $x \to -\infty$, and is unbounded as $x \to \infty$.

Bound of a Set. A number that is larger than or equal to every member of the set, is an *upper bound* of the set, and the number that is less than or equal; to every member of the set is *called lower* bound of the set. *Least upper bound axiom* : "A set of real numbers that has an upper bound, has a least upper bound." Greatest lower bound axiom: "A set of real numbers that has a lower bound, has a least lower bound."

Boundary Condition. In a differential equation the values of the variables at a certain point or information about their relationship, at a point, is used in evaluating the arbitrary constants. For example, the differential equation

$$\frac{d^2y}{dx^2} - 5\frac{dy}{dx} + 6y = 0$$

has a general solution as $y = Ae^{2x} + Be^{3x}$ where A and B are arbitrary constants. If the boundary condition are $y = 0$ and $\frac{dy}{dx} = 1$ at $x = 0$ then $A = -1, B = 1$.

Boundary Value Problems. Refers to the problems of finding a solution to give differential equation or a set of differential equations, which will meet certain specified requirements for a given set of values of the independent variables—the boundary points. Many of the problems of mathematical physics are of this type.

Branch (jump). Refers to a departure from the normal sequential execution of instructions in a computer program. Control is therefore transferred to another part of the program rather than passing in strict sequence from one instruction to the next. Refers to the branch that will be either unconditional, i.e., it will always occur, or it will be conditional, i.e., the transfer of control will depend on the result of some arithmetical or logical test.

Bromwich's Expansion Theorem.

$$\text{If} \quad u = \frac{F(p)}{f(p)} (Gt)$$

where G is a constant, the solution is

$$u = G\left\{N_0 t + N_1 + \sum \frac{F(\alpha)}{\alpha^2 f'(\alpha)} e^{\alpha t}\right\}$$

where N_0 and N_1 are defined by

$$\frac{F(p)}{f(p)} = N_0 + N_1 p + N_2 p^2 + \ldots.$$

α is any root of f(p) = 0, f(p) is the first derivative of f(p) with respect to p and the summation is overall the roots α. The solution reduces to u = 0 when t = 0.

Budan's Theorem. For a real algebraic equation of the form

$$f(x) = a_0 x^n + a_1 x^{n-1} = \ldots a_n + 0$$

where $a_0 \neq 0$

if N(x) is the number of changes of sign in the sequence of derivatives f(x), f'(x), f'' (x), ... f'' (x) where vanishing derivatives are disregarded, then the number of real roots located between two real numbers a and b > a, where a and b are not roots themselves, is either N(a)—N(b) or less than N(a)—N(b) by a positive even integer.

Buffer Store (buffer). Refers to a small area of the main store of a computer wherein information can be stored temporarily before, during, and after processing. A buffer can be used between a peripheral device and the central processor, that operate at very different speeds.

Burgers' Vector. The vector to close **Burgers' Circuit**. It is at right angles to the dislocation for a pure edge dislocation and parallel to the dislocation for a screw dislocation.

Burnside's Formula. A formula for double integration:

$$\int_{-1}^{+1}\int_{-1}^{+1} f(x, y)\,dxdy = \frac{40}{49}\{f(a,0) + f(0, a) + f(-a,0) + f(0, -a)\} + \frac{9}{49}\{f(b,b) + f(b, -b) + f(-b, b) + f(-b, -b)\}$$

where $a = \frac{\sqrt{7}}{15}$, $b = \frac{\sqrt{7}}{9}$ and $f(x, y)$ is a polynomial of degree 5 at most.

Bushel. Refers to a unit of capacity generally used for solid substances. In the UK it is equal to 8 UK gallons. In the USA it is equal to 64 US dry pints or 2150.42 cubic inches.

Byte. A unit of information in a computer consisting of a group of bits (8 in most small computers, 16 in larger ones). For some purposes a byte is equal to a character, e.g., a disk holding 250 000 bytes will hold this number of characters.

C

CAD. (Computer-aided design). The use of computers in the design of products, buildings, electronic circuits, etc. The programs take basic design criteria together with data about component sizes, tolerances, etc. Usually they display plans or three-dimensional views of the product, and these may be modified by the designer. Systems used for CAD can also list analyses of the design (e.g. safety factors, or electrical characteristics of circuits). Often, the output can be transferred directly to a computer-aided manufacturing system(CAM).

Calculation. The process of getting results from the given data by making use of mathematical procedures. Earlier the term meant a numerical result, but it is now used as a general term for any method which produces a required numerical or otherwise. *Computation* now result mean numerical calculation.

Calculator. Any device for carrying out mathematical functions (+, —, × , ÷). Originally mechanical, they were later electrically operated; purely electronic calculators, based on the digital *computer, have now replaced all others. Electronic calculators use a silicon chip integrated - circuit package, often smaller than a 2p piece, that contains the equivalent of hundreds of *transistors built into complex logic circuits, which perform the basic functions using binary arithmetic (see binary system). Multiplication and division are performed by repeated additions or subtractions. The high speed of a modern calculator enables the solution to a problem to appear on the digital dispaly apparently instantaneously. Many electronic calculators have more complex functions available (percentages, square roots, trigonometric functions, etc.) by pressing a single key and some provide a memory in which the intermediate results of a chain of calculations can be stored. Advanced types enable the user to program a sequence of functions for the calculator to perform automatically.

Calculus. The mathematical techniques, developed by Newton and *Leibniz, based on the concept of infinitely small changes in continuously varying quantities. For example, calculus is used to define the velocity of a moving body as the rate of change of its position at any instant. Velocity(v) is said to be the derivative of position (x) with respect to time (t); in calculus it is written $v = dx/dt$. In this notation dt is a vanishingly small time interval and dx is the distance the body travels in this time. If x is a known function of t, values for v at any time can be obtained by calculating the derivative of (differentiating)this function with respect to time. Similarly, the derivative of the velocity (or the second derivative of position) at any instant gives its acceleration (a), i.e. $d^2x/dt^2 = dv/dt = a$. The **differential calculus** is the system of rules for making such calculations. On a graph, the derivative of a function is the gradient of the curve at any point. A maximum or minimum value of the function can be found, as they occur when the gradient is zero. The **integral calculus** is concerned with the same process in reverse. If velocity is a continuously varying function of time, the change in position over a measurable time interval is calculated by summing the products of v and dt for each of the infinitely small intervals of time (dt) that make up the interval. The extraction of integrals follows general rules, each type of function having a corresponding integral. The integral of v between time t_1 and t_2 written.

$$\int_{t1}^{t2} v.\, dt,$$

gives the change in position in this interval. On a graph, the area between the curve of a function and the horizontal axis is the integral of the function over the specified interval.

Equations that contain derivatives are called **differential equations.** They are solved by guessing at a general form of solution and trying it in the equation, using the rules of differential calculus.

Calendar. Any of a variety of systems for the reckoning of time over an extended period. Calendars are usually based on the earth's orbital period around the sun (a year), although in some systems the moon's orbital period around the earth (a month) is taken as the basis. Since the seasons recur after each tropical *year (which contains 365.2422 days), the length of the calendar year, averaged over many years, should correspond as closely as possible to that of the tropical year. This is achieved by using leap years, which contain one more day than the usual calendar year.

In 46 BC, Julius Caesar established the so-called **Julian calendar** in which a period of three years, each of 365 days, was followed by a leap year of 366 days. The average length of the year was therefore 365.25 days. Since this was over 11 minutes longer than the tropical year, an extra day appeared about every 128 years. This discrepancy was amended by the **Gregorian calendar,** which was introduced in Roman Catholic countries in 1582 by Pope Gregory XIII, was made law in Britain and its colonies in 1752, and is now in almost world-wide use. Leap years are restricted to century years divisible by 400(e.g.1600 and 2000) and any other year divisible by four. This reduces the average length of the calendar year to a much more acceptable 365.2425 days.

Calorie or Calory. Refers to the amount of heat required to raise the temperature of one gram of water through one degree Centigrade. The calorie thus defined differs slightly for different temperatures. A standard calorie is generally defined as the amount of heat required to raise the temperature of one gram of water from 14.5° to 15.5°C. This is about the average amount required to raise the temperature of one gram of water through one degree at any point between 0° and 100°C. A more exact definition (generally accepted in the U.S.) is that one calorie equals 4.1840 absolute joules.

CAM (computer-aided manufacture). The use of computers in the organization and control of certain manufacturing processes. Examples are the control of throughput in a continuous chemical production plant, or the control of automatic machine tools in the mass-production of components (*see* robotics). Computer-aided manufacture also involves automatic stock control, scheduling, etc.

Cancel. Means (1) To divide numbers (or factors) out of the numerator and denominator of a fraction:

$$\frac{6}{8} = \frac{2 \times 3}{2 \times 4} = \frac{3}{4}.$$

the common number 2 having been *cancelled out*. (2) Two quantities of opposite sign but numerically equal are said to *cancel* when added; $2x + 3y — 2x$ reduces to $3y$, the terms $2x$ and $— 2x$ having *cancelled out*.

Cantor Set. Refers to the set of numbers in the closed interval [0, 1] that are not in the open middle third of [0, 1] and not in the open middle third of any interval obtained by separating [0, 1] into three equal intervals, then dividing each of these intervals into three equal intervals, etc. A number belongs to the Cantor set if and only if it has a representation in the ternary system of type $d_1 d_2 \ldots$, where each d_n is either 0 or 2. The Cantor set is perfect and nondense and all its points are *frontier points*. Also known as the **Cantor discontinuum,** and the **Canton ternary set.**

Cap. (1) The symbol $\cap$ used to denote the intersection of two sets. *See* intersection. (2) In connection with a sphere it refers to the smaller portion of the sphere when it is intersected by a plane other than through its centre.

Card. *Punch Card.* Refers to a rectangular piece of quality paper of size 18.73 × 8.25 cm. A punch card is the most commonly used and cheap means of feeding information and data into a computer. Information is recorded as a pattern of rectangular holes in the cards. A card has 80 columns and 12 rows wherein holes are punched. The first row is numbered 'Y', or '12' second row as 'X' or '11', and other rows are marked 0 to 9 from top to bottom. Cards are punched on a punching machine that is quite similar to a typewriter. Pressing a key results in a rectangular hole in a particular combine. Hole or combination of holes in a column stands for different characters. A single hole punched in any of 0 to 9 rows represents a digit from 0 to 9 respectively whereas an alphabet needs two holes one in 'X', 'Y' or 0 rows and second in 0-9 rows. For a special character a combination of three holes is there in a column. There is a provision for getting the characters printed also at the top of the card.

Cartesian Coordinates. Refers to a method of defining the position of a point by its distance from a fixed point (origin) in the direction of two or more straight lines. On a flat surface, two straight lines known as the x-axis and the y-axis, form the basis of a two-dimensional Cartesian coordinate system. The point where they cross in the origin. An imaginary grid is formed by lines parallel to the axes and one unit length apart. The point (2, 3), for example, is the point at which the line parallel to the y-axis two units in the direction of the x-axis, crosses the line parallel to the x-axis three units in the direction of the y-axis. Mostly the x-axis is horizontal and the y-axis is perpendicular to it. These are called ***rectangular coordinates***. If the axes are not a right angles they are ***oblique coordinates.***

In three dimensions, a third, z-axis, is added to define the height or depth of a point. The coordinates of a point are then three numbers (x, y, z). A right-handed system is one for which if the thumb of the right hand points along the x-axis, the fingers of the hand fold in the direction in which the y-axis would have to rotate to point in the same direction as the z-axis. A left-handed, system is the mirror image of this. In a rectangular system, all three axes are mutually at right-angles.

Cassini's Ovals. Curves giving the locus of points having the product of their distances from (—a, 0) and (a,0) and c^2 and having the form

$$(x^2 + y^2 + a^2)^2 - 4a^2x^2 = c^4$$

Castigliano's Theorems. I. If U is the strain energy (see **Clapeyron's Theorem**) when a body is deformed, s_i are the displacement components and F_i are the corresponding force components, then

$$s_i = \frac{\partial U}{\partial F_i}$$

II. If the prescribed values of the displacement components are denotes s'_i then the unprescribed forces F_i are such that $(U - \Sigma s'_i F_i)$is a minimum.

Catenary. Refers to the plane curve of a flexible uniform line suspended from two points. For example, an empty washing line attached to two-poles and hanging freely between them follows a catenary. The catenary is symmetrical about an axis perpendicular to the line joining the two points of suspension. In Cartesian coordinates, the equation of a catenary which has its axis of symmetry lying along the y-axis, and that crosses the y-axis at y = a, is given by

$$y = (a/2)\,(e^{x/a} + e^{-x/a}).$$

Catenoid. Refers to the curved surface obtained by rotating a catenary about its axis of symmetry.

Cauchy—Riemann Equations. C-R *Equations*. A necessary condition for a function of complex variables to be analytic. A function f(z) = u + iv, [where z = x + iy, u = u (x, y) and v = (x,y)] is analytic it

$$\frac{\partial u}{\partial x} = \frac{\partial v}{\partial y} = \frac{\partial u}{\partial y} = -\frac{\partial v}{\partial x}.$$

These equations are known as Cauchy — Riemann equations. The polar form of these equations are

$$\frac{\partial u}{\partial r} = \frac{1}{r}\frac{\partial v}{\partial \theta}, \quad \frac{\partial v}{\partial r} = -\frac{1}{r}\frac{\partial u}{\partial \theta}.$$

Cauchy's Condensation Test. If f (n) is a positive decreasing function of n, and a is any positive integer greater than 1, the two series $\Sigma f(n)$ and $\Sigma a^n f(a^n)$ are both convergent or both divergent.

Cauchy's Convergence Criterion. A sequence (u_n) converges if and only if, for each $\varepsilon > 0$, there exists a number N, such that $| u_p — u_q | < \varepsilon_0$, for p, q > N. This criterion has the advantages that one need not know the limit to decide the convergence.

Cauchy's Dispersion Formula.

$$n = A + \frac{B}{\lambda^2} + \frac{C}{\lambda^4}$$

where n is the refractive index and A, B and C are constants. As A, B and C are determined from observations of n at widely different wavelengths, the formula represents the dispersion of most substances with considerable accuracy, although the theory used by Cauchy, the elastic solid theory of light, is not correct.

Cauchy's Frequency Distribution. A frequency distribution for the variable x given by,

$$f(x) = \frac{1}{\pi a} \frac{1}{1 + [(x — \lambda)/a]^2}$$

where the distribution is symmetric about λ, which is the median but not the mean. α is a measure of the spread of the distribution.

Cauchy's Generalized Theorem. If f(x) and g(x) are continuous in [a, b] and differentiable in (a, b), then there exists a point S in (a, b) such that

$$\frac{f(b) — f(a)}{g(b) — g(a)} = \frac{f'(S)}{g'(S)},\quad g(a)\, g \neq (b) \text{ and } f'(x) \text{ and } g'(x)$$

are not zero simultaneously.

Cauchy's Integral Formula. If f(z) is analytic inside and on a closed contour C and $z = z_0$ is a point inside C then.

$$\frac{1}{2\pi i} \int_C \frac{f(z)\, dz}{z — z_0} = f(z_0),$$

$$\frac{1}{2\pi i}\int_C \frac{f(z)\,dz}{(z - z_0)^2} = f(z_0).$$ Continuing the process

$$\frac{n!}{2\pi i}\int_C \frac{f(z)\,dz}{(z - z_0)^{n+1}} = f^{(n)}(z_0)$$

Cauchy's Mean Value Theorem. If f(x) and $\phi(x)$ are continuous in the range $a \leq x \leq b$, and f (x) and ϕ' (x) exist and do not vanish simultaneously in the range $a < x < b$, then there exists a point c such that $a < c < b$ where

$$\frac{f(b) - f(a)}{\phi(v) - \phi(b)} = \frac{f'(c)}{\phi'(c)}$$

Cauchy's Residue Theorem. If f(z) represents analytic within and on a closed curved C except at a finite number of points $z = z_1$, z_2 ,... z_n inside C then

$$\int_C f(z)\,dz = 2\pi i \sum_{r=1}^{n} \text{Res } f(zr)$$

Cavalieri's Theorem. If two solids have equal altitudes and all plane sections parallel to their bases and at equal distances from their bases have equal areas, then the solids have the same volume.

Cayley Algebra. Refers to the set of symbols of type A +Be, where A and B are quaternions and addition and multiplication are difined by

$$(A + Be) + (C + De) = (A + C) + (B + D)e,$$

$$(A + Be)(C + De) = (AC - BD) + (AD + BC)e,$$

with $\bar{C}$ and $\bar{D}$ the conjugates of the quaternions C and D. Except for multiplication not being associative, the Cayley algebra satisfies all axioms for a division algebra having unit element. As a vector space over the field of real numbers, it is of dimension 8 having the basis (1, i, j, k, e, ie, je, ke,). The definition of multiplication means that $e^2 = - 1$, $ie = - ei$, je = — ej, ke = — ek, but that (ij)e = ke and i(je) = — ke. The members of the Cayley algebra are **Cayley numbers.**

Cayley-Klein Parameters. Quantities used to express the orientation of a rigid body. If the parametrers are represented by α, β, γ and δ, then they are related to the **Eulerian Angles** by

$$\alpha = \exp[j(\chi + \phi)/2] \cos \theta/2$$

$$\beta = j \exp[j(\chi - \phi)/2] \sin \theta/2$$

$$\gamma = j \exp[-j(\chi - \phi)/2] \sin \theta/2$$

$$\delta = \exp[-j(\chi + \phi)/2 \cos \theta/2$$

Cayley Numbers. These have eight base elements : 1, e_1, ... e_7,

with $e_i^2 = -1 \quad e_ie_j = -e_je_i$

and $e_1e_2 = e_3 \quad e_1e_4 = e_5 \quad e_1e_6 = e_7 \quad e_2e_5 = e_7$.

$$e_2e_4 = -e_6 \quad e_3e_4 = e_7 \quad e_3e_5 = e_6.$$

All other products of base elements are obtained by the further rule that $e_ie_k = e_j$ implies that $e_ke_j = e_i$ and $e_je_i = e_k$.

Cayley's Theorem. Every finite group is isomorphic with a certain group of permutations. Thus the study of finite groups may be reduced to a study of the permutation groups, and the theorem has important application to symmetry studies. Two groups G (= g_1, g_2, ...) and F = (f_1, f_2, ...) are called isomorphic if a one-to-one correspondence can be established between their elements and if their multiplication tables coincide, i.e. if

$$g_i \longleftrightarrow f_i \quad g_j \longleftrightarrow f_j \quad g_ig_j \longleftrightarrow f_if_j$$

Celcius Grade (Symbol °C). Refers to a scale of temperature that is named after the Swedish Physicist Celsius who based on the freezing temperature of water(0°C) and the boiling point of water(100°C) is fixed extremes at normal atmospheric pressure.

The interval between these two temperatures is divided

into 100 equal parts, each part is called 1°C.

Central Force. Refers to a force that acts on any affected body along a line to an origin. For example, the motion of electric forces between charged particles are central whereas frictional forces are not.

Central Processor (central processing unit, CPU). Refers to a highly complex electronic device that is the nerve centre of a computer. It is made up of the *control unit* and the *arithmetic and logic unit* (ALU). Also sometimes considered part of the central processor is the main store; or memory, where a program or a section of a program is stored in binary form.

The control unit supervises all activities within the computer, interpreting the instructions that make up the program. Each instruction is automatically brought, in turn, from the main store and kept temporarily in a small store known as a *register*. Electronic circuits analyse the instruction and determine the operation that are to be carried out and the exact location or locations in store of the data on which the operation is to be performed. The operation is actually performed by the ALU, again using electronic circuitry and a set of registers. It could be an arithmetical calculation, like the addition of two numbers, or a logical operation, like selecting or comparing data. This process of fetching, analysing, and executing instructions is repeated in the required order until an instruction to stop is executed.

The size of central processors has diminished considerably with advances in technology. It is now possible to form a central processor on a single silicon chip a few millimetres square in area, or on a small number of chips. This tiny device is called a microprocessor.

Central Projection (conical projection). Refers to geometrical transformation wherein a straight line form a point (called the *centre of projection*) to each point in the figure is continued to the point at which it passes through the second (image) plane. These points form the image of the original figure. When a photographic image is obtained from a film using an enlarger, this is the kind of projection that occurs. The light source is at the centre of projection, the light rays are the straight lines, the film is the first plane, and the screen or point is the second. In this case the two planes are generally parallel, but this is not always so in central projection.

Centre of Buoyancy (for an object in a fluid). Refers to the centre of mass of the displaced volume of fluid. For a floating object to be stable the centre of mass of the object must lie below the centre of buoyancy; when the object is an equilibrium, the two lie on a vertical line.

Centre of Mass. Refers to the point in a body where the entire mass may be concentrated, theoretically, to produce the same effect as when distributed. The term is slightly different than centre of gravity. In uniform gravitational field centre of mass is the same as centre of gravity. The centre of mass coincides with the centre of symmetry if a symmetrical body has a uniform density throughout.

If $(x_1, y_1, z_1), (x_2, y_2, z_2), \ldots, (x_n, y_n, z_n)$ are coordinates of masses $m_1, m_2, \ldots, m_n$ then the centre of mass (x, y, z), is given by

$$x = \frac{\sum_{r=1}^{n} m_r x_r}{\sum_{r=1}^{n} m_r}, \quad y = \frac{\sum_{r=1}^{n} m_r y_r}{\sum_{r=1}^{n} m_r}, \quad \bar{z} = \frac{\sum_{r=1}^{n} m_r z_r}{\sum_{r=1}^{n} m_r},$$

Centripetal Force. Refers to a force that causes an object to move in a curved path rather than continuing in a straight line. The force is provided by :

— the tension of the string, for an object whirled on the end of a string;

— gravity, for an object in orbit round a planet;

— electric force, for an electron in the shell of an atom.

The centripetal force for an object of mass m with constant speed v and path radius r is mv^2/r, or $m\omega^2 r$, where ω is angular speed. A body moving in a curved path has an acceleration as the direction of the velocity changes, even though the magnitude of the velocity may remains constant. This acceleration, that is directed towards the centre of the curve, is the *centripetal acceleration*. It is given by v^2/r^2 or wr.

Centroid. Refers to the point in a plane figure or a solid where the centre of mass would be if the body has uniform density throughout. Centroid of a symmetrical figure wil be the centre of symmetry. For example, the centroid of a triangle is the point of intersection of its medians, centroid of circle or a sphere is its centre.

Cesaro's Summation Formula. A method of obtaining a sum of a series Σx_i which may not necessarily be convergent. If s_1, s_2, ... s_n are partial sums of the series, then the sum is given by

$$\lim_{n \to \infty} \frac{s_1 + s_2 + \ldots s_n}{n}$$

For example, for the series 1 — 1 + 1 — 1 + ..., the method obtains a value of 0.5.

C.G.S.System. Refers to a system of physical units, wherein unit of mass is *gram*, unit of length is *centimetre* and unit of time is *second*.

Chain Rule. Refers to a rule used for differentiating a function of a function. If z is a function of y and y is a functions of x then

$$\frac{dz}{dx} = \frac{dz}{dy} \cdot \frac{dy}{dx}.$$

The chain rule can be extended to functions of several variables as well.

Chance. Same as probability. Has considerable popular, but little techincal, usage.

Character-Finite Character. A *collection* A of *sets* of finite character if A contains any set all of whose finite subsets belong to A and each finite subset of a member of A belongs to A. ***A propertry of subsets*** of a set is of finite character if a subset S has the property if and only if each nonempty finite subset of S has the property. The property of being ***simply ordered*** is of finite character, while the property of being ***well ordered*** is not. If a property is of finite character, then the collection of all sets having this property is of finite character. If a collection A of sets is of finite character, then the property of belonging to A is of finite character.

Characteristic. Refers to the integral part of the logarithm of a number determined by the place value of the digit with the highest place value. Thus, unit numbers have characteristic 0, tens and tenths 1 and — 1 respectively, hundreds and hundredths 2 and — 2 respectively, thousands and thoudandths 3 and — 3 respectively etc.

Chebyshev's Approximation. An arbitrary function can be expanded in a given finite range as a power series (e.g. **Taylor's Series**). An expansion in terms of **Chebyshev Polynomials** will ensure that the value of the deviation of the resulting polynomial from the original function will be not greater than a given pre-assigned number.

Chebyshev's Inequality. If x is a random variable of mean value $\bar{x}$ and σ is the standard deviation, then there is an upper limit to the probability that the absolute deviation of x from $\bar{x}$ is greater than or equal to a given value a(> 0), and this probability limit is given by

$$P[\,|x-\bar{x}|\geq a] \leq \frac{\sigma^2}{a^2}$$

Chi-square Distribution (χ^2 distribution). Refers to the distribution of the sum of the squares of random variables having standard normal distributions. For examples, if $x_1, x_2, \ldots x_n$ are independent and all standard normal, then

$$\chi^2 = \Sigma x^2 i$$

has a chi-square distribution having n degrees of freedom, written χ_n^2. The mean and variance are n and 2n respectively.

The values $\chi_n^2(\propto)$ for which $P(X^2 < \chi_n^2(\propto)) = \propto$ are tabulated for various values of n.

Chi-square Test. Refers to a measure of how well a theoretical probability distribution fits a set of data For i = 1, 2, .. m the value x_i, occurs 0_i, times in the data and the theory predicts that it will occur e_i, times. Provided $e_i \geq 5$ for all values of i(otherwise values must be combined), then

$$\chi^2 = \Sigma\,(0_i - e_i)^2/e_i$$

has a chi-square distribution having n degrees of freedom.

Circle. A curve defined as the locus of all the points lying in a plane at a certain distance, called the radius, from a fixed point, called the centre. Its diameter is any straight line joining two points on the circle and passing through the centre. The ratio of the distance around any circle (the circumference) to its diameter is equal to the number π. The area of a circle, radius r, is πr^2.

Circum-centre Circumcentre of a Triangle. Refers to the centre of the circumscribed circle (the circle passing through the three vertices of the triangle); the point of intersection of the perpendicular bisectors of the sides; point O in the figure.

Circum Circle. Same as circumscribed circle.

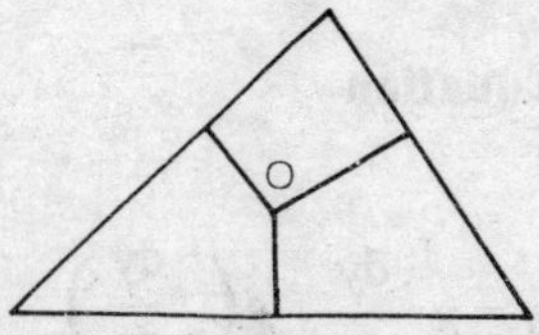

Fig. C-1.

Circumscribed. Refers to a geometric figure which is drawn around and enclosing another geometrical figure. For example, in a square, a circle can be drawn through the vertices. This is called the circumscribed circle, and the square is called the *inscribed* square of the circle. Similarly, a regular polyhedron might have a circumscribed sphere, and a rectangular pyramid a circumscribed one.

Cissoid. Refers to a curve having polar equation as

$$r = \frac{2a \sin^2 \theta}{\cos \theta}$$

and cartesian equation as

$$y^2 (2a - x) = x^3.$$

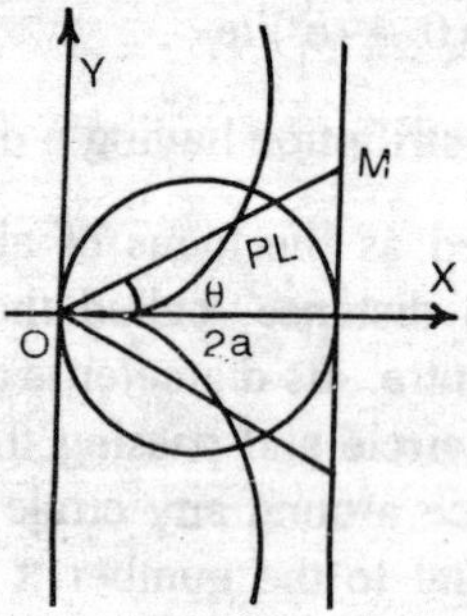

Fig.C-2.

Let O be a fixed point on the circumference of circle of radius a. Draw any secant through O cutting the circle at L and the tangent at the extremity of the diameter through O, at M. If the P is a point on OM such that OP = LM then the locus of the point P is the required cissoid with x = 2a as asymptote.

Clairaut's Differential Equation

If

$$y = x \frac{dy}{dx} + f\left(\frac{dy}{dx}\right)$$

its general solution is of the form y = cx + f(c).

The singular solution is obtained by eliminating dy/dx between

$$y = x\frac{dy}{dx} + f\left(\frac{dy}{dx}\right) \text{ and } x + \frac{d}{dx}f\left(\frac{dy}{dx}\right) = 0$$

Clairaut's Theorem. If g_e and g_λ are the values of the acceleration due to gravity at the earth's equator and latitude λ respectively, and m is the ratio of the centrifugal acceleration to g_e, then

$$g_\lambda = g_e\,[1 + (5/2m - e)\sin^2 \lambda],$$

where e = ellipticity = $(r_1 - r_2)/r_1$ and r_1 and r_2 are the equatorial, and polar radii. Bouguer and Faye have given correction rules for local variation of altitude.

Class Limits. Means limits of different classes in a frequency distribution. If the person in a locality are classified into the age groups 0 — 5 years, 5 — 10 years, etc., then it should be clearly decided as to which group as child will be put, who is just 5 years old. Usually it is put in the class interval 5—10. As a general rule in a class interval the upper limit is excluded. For example, the class interval 0—15, contains the quantities whose value is less than 15.

Clockwise. Means rotating in the direction as the hands of a clock. The anti-clockwise rotation is taken as positive and the clockwise as negative.

Clebsch-Gordan Coefficients. If D_i and D_k are irreducible representations

$$D_i \times D_j = \sum_k C^k_{ij} D_k$$

then C^k_{ij} are Clebsch-Gordan coefficients and the above expression, called the Clebsch-Gordan series, given the number of times the irreducible representation D_k occurs in the **Kronecker Product** of D_i and D_j. The series permits the determination of selection rules for transitions between energy levels.

Closed Curve (closed contour). Refers to a continuous curve whose end points coincide. For example, a circle, an ellipse, etc, whose forms a complete loop and has no end points. A *simple closed curve* is one that does not intersect itself.

Closed Interval. Refers to an interval that includes the two end points. For example, set of all real numbers greater than or equal to 0 and less than or equal to 1, denoted by [0, 1].

Closed Set. Refers to a set that includes its all limit points. For example, set of all rational numbers x such that $2 \leq x \leq 3$, set of all points lying inside and on a circle.

Closed Surface. Refers to a continuous surface that does not have any boundary lines or boundary curves. For example, a sphere polyhedra, ellipsoid.

Cockran's Theorem. If X_i (e = 1, ..., n) are independently and normally distributed random variables with zero mean and unit variance, and if $q_1, q_2, ..., q_k$ are k quadratric forms in the variables X_i having ranks $r_1, r_2, ..., r_k$ and $\sum_{j=1}^{k} q_j = \sum_{i=1}^{n} X_i^2$, then a necessary and sufficient condition that q_j be cach independently distributed with the X^2 distribution with r_j degrees of freedom is that $\sum_{j=1}^{k} r_j = n$. On the basis of this theorem it follows that if $\{X_1, ..., X_n\}$ is a random sample from a normal distribution with mean u and variance σ^2, then $\sum_{i=1}^{n} \frac{(X_i - \bar{X})^2}{\sigma^2}$ is distri—buted as X^2 having n — 1 degrees of freedom, where $\bar{X}$ is the mean of the sample. This theorem is quite useful, for example, in establishing the independence of the mean and sum of squares of deviations around the mean in random samples from a normal population.

Cobol (common business oriented language). An international computer-programming language, developed in the USA and used to express problems in commerce. A high-level language, it can describe business procedures in the form of readable English statements that are sufficiently standardized for use in a computer.

Coefficient of Correlation. In the given data if the variables x and y are related to each other, then the coefficient of correlation denoted by ρ or r is the quotient of $\Sigma(xy)$ and $\sqrt{\Sigma x^2 \Sigma y^2}$. Hence, the measure of correlation between the variables x and y is given by ρ (or r) $= \dfrac{\Sigma(xy)}{n\sigma_x\sigma_y}$, where σ_x and σ_y are the standard deviations of x and y respectively and n is the number of pairs of values of x and y. The coefficient of correlation always lies between — 1 and 1.

Combination. If out of n available things, r things are to be selected at random, it can be done in $\dfrac{n!}{r!\,(n-r)!}$ ways and is denoted by ^nC_r where the letter C stands for combinations. Other notations used are $\binom{n}{r}$, C (n, r), C_H here

$$^nC_0 = {}^nC_n = 1$$

and $^nC_0 + {}^nC_1 + {}^nC_2 + ... + {}^nC_n = 2^n$.

Commensurable. Means measurable in terms of same units. For example, a 30 centimeter length is commensurable with 1 meter length as both can be measured in centimetres, but the length of a side and that of a diagonal of a square are not commensurable.

Common Denominator. A positive integer that is divisible by the denominators of two or more fractions. For example, 4 and 8 both are common denominators of the fractions 1/2 and 1/4 of which 4 is the lowest (least) common denominator (LCD).

Common Multiple. Refers to a whole number that is a multiple of each of a group of numbers. For example, 100 is a common multiple, of 5, 25 and 50. The *lowest* (or *least*) *common multiple* (LCM) is the smallest number that is a common multiple : in this case it is 50.

Common Tangent. Refers to a single line which forms a tangent to two or more separate curves. Also refers to the length of the line joining the two tangential points.

Commutative. Represents an operation that is independent of the order of combination. A binary operation is commutative if a b = b a for all values of a and b. In ordinary arithmetic, multiplication and addition are commutative operations. This is known as the *commutative law of multiplication* and *the commutative law of addition*. Substraction and division are not commutative operations.

Commuting, Commuting Obligations. Exchanging one set of obligations to pay a certain sum (or sums) at different times for another to pay in accordance with some other plan. The common data of comparison at which the two sets are equivalent (equal in value at that time) is the focal date.

Compact. A topological space E being compact (originally, *bicompact*) means that it has the property that, for any union of open sets which contains E, there is a finite number of these open sets whose union contains E. A space E is compact if and only if each collection of closed sets contained in E has a nonempty intersection if it has the *finite intersection property*. e.g. All closed subsets of a compact space are compact; all compact subsets of a Hausdorff topological space are closed. A *locally compact* is a space E having the property that each point of E has a neighbourhood which is contained in a compact subset of E. The set {0, 1, 1/2, 1/3, 1/4,......} is compact. The set R of real numbers is not compact, since R is contained in the union of all open intervals of unit length, but R is not contained in any finite union of such intervals. A countably *compact space* is a space E having the property that, for any union of a countable number of open sets which contains E, there is a finite number of those open sets whose union contains E. A space E is countably compact if and only if each sequence in E has an accumulation point in E. A *sequentially compact* (*originally, compact*) space is a space E having the property that each sequence in E contains a subsequence which converges to a point in E. For a space to have the *Bolzano-Weierstress* property means that each infinite subset has at least one accumulation point. For a Lindel of space (and therefore for a metric space), compactness, countable compactness, sequential compactness, and the Bolzano-Weierstrass property are equivalent. All compact spaces are countably compact, and all countably compact spaces have the Bolzano-Weierstrass property. All T_{-1} spaces with the Bolzano-Weierstrass property are countable compact. All sequential compact spaces are countably compact, and all countably compact spaces satisfying the first countability axiom are sequentially compact.

Compasses. Refers to an instrument used for drawing circles. It consists of two rigid arms joined by a hinge. At one end is sharp point, that is placed at the centre of the circle. At the other end is a pencil or other marker, that traces out the circumference when the compasses are pivoted around the sharp point. In a *beam compass,* used for drawing large circles, the sharp point and the marker are connected to opposite ends of a horizontal beam.

Comparison Test. Refers to a test for convergence and divergence of an infinite series. If from and after certain term absolute value of each term of the series under test is less than or equal to the corresponding term of some convergent series, the given series is called a convergent. Similarly, if from and after certain term absolute value of each term of the series under test is greater than or equal to corresponding term of a divergent series, the given series is called a divergent.

In other words, if $\underset{n \to \infty}{\text{Lt.}} \frac{u_n}{v_n}$, is non-zero and finite then the series Σu_n behaves like the known series Σv_n does. The series Σv_n is generally taken as $\Sigma \frac{1}{n^p}$ that is already known to be convergent if $p > 1$ and divergent if $p \leq 1$.

Complementary Angle. Refers to a pair of angles whose sum is a right angle $\left(90° \text{ or } \frac{\pi}{2}\right)$. One angle of the pair is the complement of the other.

Completing the Square. A method generally used for factorising a quadratic expression or solving a quadratic equation. The method involves adding of a quantity to a given quantity so that the sum becomes a perfect square. For example, the quadratic equation $ax^2 + bx + c = 0$ can be solved as

$$x^2 + \frac{b}{2a} + x + \frac{c}{a} = 0 \text{ or } \left(x + \frac{b}{2a}\right)^2 - \frac{b^2}{4a^2} + \frac{a}{c} = 0$$

$$\text{or } \left(x + \frac{b}{2a}\right) - \left(\frac{\sqrt{b^2 - 4ac}}{2a}\right) = 0$$

$$\text{or } \left[x + \frac{b}{2a} + \frac{\sqrt{b^2 - 4ac}}{2a}\right]\left[x + \frac{b}{2a} - \frac{\sqrt{b^2 - 4ac}}{2a}\right] = 0$$

which gives $x = \dfrac{-b \pm \sqrt{b^2 - 4ac}}{2a}$.

Complex Number. A number oif the form $z = x + iy$ where x and y are real numbers and $i = \sqrt{-1}$. x is called the real part of the complex number z, written as Re (z) and y the imaginary part of z, written as I_m (z) 0 is also a complex number since $0 = 0 + 0.i$. To every complex number z there appears $Z = x - iy$ and $-z = -x - iy$.

Addition and subtraction

$$(a_1 + ib_1) \pm (a_2 + ib_2) = (a_1 \pm a_2) + i\,(b_1 \pm b_2)$$

Multiplication $\quad (a_1 + ib_1)(a_2 + ib_2)$

$$= (a_1 a_2 - b_1 b_2) + i\,(a_1 b_2 + a_2 b_1)$$

Division

$$\frac{a_1 + ib_1}{a_2 + bi_2} = \frac{(a_1 + ib_1)(a_2 - ib_2)}{(a_2 + ib_2)(a_2 - ib_2)}$$

$$= \frac{(a_1 a_2 + b_1 b_2) + i\,(a_2 b_1 - a_1 b_2)}{a_2^{\,2} + b_2^{\,2}}$$

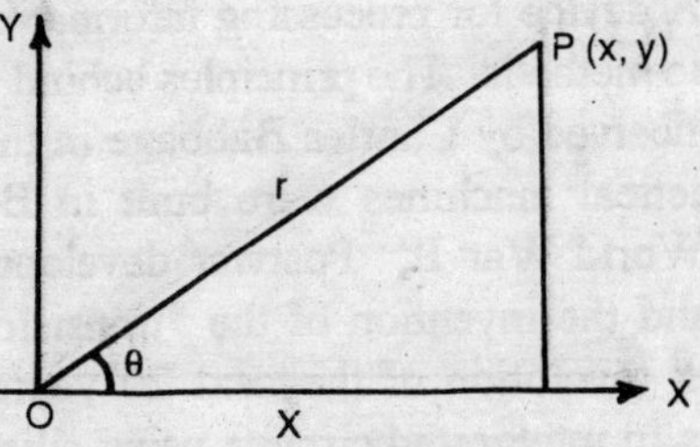

Fig. C-3.

A complex number can always be shown by an Argand diagram. As shown in the figure the complex number ($z = x + iy$) is given by the point (x, y). The horizontal axis is called real axis measuring real part of z, and the vertical axis the imaginary axis that measures the coefficient of i in z i.e. the imaginary part of z. Origin O represents 0 ($0 + i0$). z may also be written in trigonometric from as $re^{i\theta} = r(\cos\theta + i\sin\theta)$ so that $x = r\cos\theta$, $y = r \sin, \theta$, r is called the modulus of z and is denoted by $| z | + \sqrt{x^2 + y^2}$ where θ is known as the argument or amplitude of z and is denoted by Arg. $(z) = \tan^{-1}(y/n)$.

Component Vectors. Refers to the components of a given vector (like a force or velocity) are two or more vectors having the same effect as the given vector. Or the given vector is the resultant of the components. Any vector has an infinite number of sets of components. Some sets are more useful than others in a given case particularly pairs at 90°. The component of a given vector (V) in a given direction is the projection of the vector in that direction; i.e. $V\cos\theta$ where θ is the angle between the vector and the direction.

Compound Interest. When the interest in each period is added to the original sum as it is earned. Thus the principal, and therefore the interest on it, increases year by year. If P is the principal (the original amount of money invested), R percent the interest rate per annum, and n and the number of interest periods, then the compound interest is

$$P(1 + R/00)^n$$

Computer. A device for processing information at high speeds, by electronic methods. The principles behind the modern computer were conceived by Charles Babbage in the 19th century. The first practical machines were built in Britain and the USA during World War II. Postwar developments in information theory and the invention of the *transistor made possible the computer revolution of the next 20 years. Since the 1970s, advances in intetgrated circuits using silicon chips have led to the development of *minicomputers and then microcomputers.

A **digital computer** processes information in the form of groups of binary numbers, which are represented by the on and off positions of electronic switches. The sequence of operations it performs on this information is controlled by a *program, and the suite of programs that enables a computer to perform useful functions is called its software. The physical equipment, or **hardware,** of a computer generally has three main components: the Central Processing Unit (CPU), main memory, and peripheral devices that enable information to be fed into the machine, display it in a readable form, or act as auxiliary memory. Input, formerly by punched tape or cards, is now by keyboard at a Visual Display Unit (VDU) or by methods of automatic data caputre, such as Optical Character Recognition (OCR). Output is by printout on paper, VDU, or magnetic disc or tape. New developments, such as the use of compact discs as large permanent memories and the development of CPUs that work much faster by using optical rather than electrical impulses, promise to increase the power of computers. The **transputer,** which harnesses several micro-processors in parallel, seems to offer microcomputer users the computing speed that is currently only available on large mainframe computers.

An **analog computer** is a device that deals with continuously varying physical quantities, such as current or voltage. They are used mainly for simulation or monitoring and controlling continuous processes in industry or scientific research. Sometimes they act as peripheral devices for a digital computer (a combination known as a **hybrid computer**).

Computer Generations. The first generation of computers, developed in the early 1950s, used valves. The second, in the late 1950s, were based on solid-state components, while the third, in the mid-1960s, had integrated circuits. In the mid-1970s, the fourth generation appeared using large-scale integration technology. The fifth generation, to be revealed in the 1990s, will have multiple parallel processors on a single chip or chip set, using Very Large Scale Integration (VLSI). Such computers will communicate with each other in networks and use artificial intelligence, voice recognition and speech synthesis.

Computer Modelling. Refers to the development of a description or mathematical representation, (i.e., *a model*) of a complicated process or system, with the help of a computer. This model can then be used to study the behaviours or control of the process or system by changing the conditions in it, again with the help of a computer.

Computer Virus. An unauthorized computer program that exploits weaknesses in a computer's operating system to insert itself into the system; it then propagates itself to other machines via networks or disks. When activated it interferes with the machine's opertation, and in some cases destroys data held on the machine. Often written as jokes, computer viruses are an increasingly serious menace.

Concave. Means curved inwards. For example, the inner surface of a hollow sphere is concave. Similarly in two dimensions, the inside edge of the circumference of a circle is concave. A *concave polygon* is a polygon which has one (or more) interior angles greater than 180°.

Condition. In logic refers to a proposition or statement, P, which is required to be true so that another proposition Q be also true. If P is *necessary condition* then Q could not be true unless P is true. If P is a *sufficient condition,* then whenever P is true Q is also true, but not vice versa. For example, for a quadrilateral to be a rectangle the necessary condition is that two of its sides be parallel, but this is not a sufficient condition. A sufficient condition for a quadrilateral to be a rhombus is that all its sides have a length of 5 centimeters, but this is not a necessary condition. For a rectangle to be a sqare it is both a necessary and a sufficient condition that all its sides are of equal length.

In formal terms, if P is a necessary condition for Q, then $Q \rightarrow P$. If P is sufficient condition, then $P \rightarrow Q$. If P is necessary and sufficient condition for Q then $P \equiv Q$.

Cone. Refers to a solid whose base is a plane closed curve also called *Guiding curve* and a point outside the plane, called *vertex*. A line segment joining the vertex to a point on the plane curve, generates the curved surface of the cone when the point moves on the guiding curve, the line is known as the *generator* and the guiding curve is known as the *directrix* of the cone. Any line segment from the vertex to the directrix is also known as *element* of the cone. If the directrix is a circle the cone is a *circular* cone and the line joining the vertex to this centre is the *axis* of the cone. If the axis is perpendicular to the base, the cone is a *right circular cone,* otherwise *oblique* cone.

The volume of right circular cone is $1/3\ \pi r^2 h$ and the area of the curved surface is πrl, where r is the radius of the base, *h* the height, and *l* the slant height, i.e., the length of an element.

Confocal Conics. Refers to conic sections with the same foci.

For example, the equation $\dfrac{x^2}{a^2 - k} + \dfrac{y^2}{b^2 - k} = 1$ stands for a family of confocal ellipses if $a^2 > b^2 > k$ and a family of confocal hyperbolas if $a^2 > k > b^2$ and the equation $y^2 = 4a\ (x - a)$ stands for confocal parabolas having origin as focus and latus rectum 4a for different values of a.

Congruent. Refers to two figures which are identical in shape and size. The figures may be in a plane or in space. The two plane figures will be congruent if one figure can be brought by rigid motion to other figure in order to fit into the area occupied by each other, coinciding exactly. For example, two circles are congruent if they have same radius. Two triangles are congruent if they satisfy any one of the following conditions:

(i) Two sides and inclued angle of one triangle are equal to the corresponding two sides and included angle of the otherr (SAS rule).

(ii) Two angles and the included side of one triangle are equal to the corresponding two angles and included side of the other (ASA rule).

(iii) Three sides of one triangle are equal to the corresponding three sides of the other (SSS rule).

Two three dimensional figures are congruent if one can be made to coincide with the other in space by rigid motion.

Conic (Conic Section). Refers to a plane curve defined as the locus of a point that moves in such a way that the ratio of its distances from a fixed point (focus) and a fixed line (*directrix*) is constant. This constant ratio is known as *eccentricity* of the conic. Conics are of three types. The conic having eccentricity equal to 1, are known as *parabolas;* eccentricity less than 1 *ellipse;* eccentricity greater than *1 hyperbola.* While a parabola has one focus and one directrix, the ellipse and hyperbola have two foci and two directrices each.

Conical Surface. In the *cone* if the generator is extended beyond the vertex and the directrix, then the surface generated by this generator is known as the conical surface. This is made up of *twin nappes.* If the directrix is the circumference of a circle and the axis of the surface is at right angles to the plane of the circle, surface is known as *right circular conical surface*. Plane sections of such a surface are called *conic sections.*

Conjugate Complex Numbers. Refers to two complex numbers of the form $x + iy$ and $x — iy$, which when multiplied together have a real product $x^2 + y^2$. If $z = x + iy$, the complex conjugate of z is $z = x — iy$.

Conjunction Symbol: V. In logic refers to the relationship between two or more propositions or statements. The conjunction of P and Q is true when P is true and Q is true, and false otherwise. The truth table definition for conjunction is given in the illustration below.

PQ	P ∧Q
TT	T
TF	F
FT	F
FF	F

conjunction

Connectivity. Refers to the number of cuts needed to break a shape in two parts. For example, a rectangle, a circle, and a sphere, all have a connectivity of one. A flat disc having a hole in it or a torus has a connectivity of two.

Consequent. In logic refers to the second part of a conditional statement : a proposition or statement which is said to follow from or be implied by, another. For example, in the statement if a man is happy then Amit is happy', Amit is happy' is the consequent.

Conservation Law. Refers to a law which states that the total value of some physical quantity is conserved (i.e., remains constant) throughout any changes in a closed system. The conservation laws as applied in mechanics are the laws of constant mass, constant energy, constant linear momentum, and constant angular momentum.

Conservation of Mass and Energy. Refers to the law that the total energy (rest mass energy + kinetic energy + potential energy) of a closed system is constant. In most chemical and physical interactions the mass change is very small, so that the measurable rest-mass energy does not change (it is regarded as 'passive'). The law then becomes the classical *law of conservation of energy*. In practice, the inclusion of mass in the calculation is necessary only in the case of nuclear changes or systems involving high velocities.

Conservative Force. Refers to a force such that, if it moves an object between two points, the energy transfer (work done)is independent of the path between the points. It must then be true that if a conservative force moves an object in a closed path (back to the starting point), the energy change is zero, e.g., Gravitation.

Consistent. (Of a theory, system, or set of propositions). Means giving rise to no contradictions. Arithmetic, is a consistent logical system because non of its axioms nor any of the theorems that are derived from them by the rules, are contradictory.

Consistent Equations. Refers to a set of equations which can be satisfied by at least one set of values for the variables. For example, the equations $x + y = 2$ and $x + 4y = 6$ are satisfied by $x = 2/3$ and $y = 4/3$ and they are thus consistent. The equations $x + y = 4$ and $x + y = 9$ are inconsistent.

Constant. Refers to a quantity which does not change its value in a general relationship between variables. For example, in the equation $y = 2x + 3$, where x land y are variables, the numbers 2 and 3 are constants. They are *absolute constant* in this case as their values never change. Sometimes a constant can take any one of a number of values in different applications of a general formula. In the general quadratic equation.

$$ax^2 + bx + c = 0$$

a, b and c are arbitrary constants as no values are specified for them. An indefinite integral includes an arbitrary constant (the *constant of integration*) because, for a function f(x) the integral with respect to x has the form

$$\int f(x)\, dx + c$$

where the value of the constant c depends on the limits chosen.

Constant Angular Momemtum, Law of. (Law of conservation of angular momentum). It states that the total angular momentum of a system cannot change unless a net outside torque acts on the system.

Continued Fraction. An integer and a fraction, the denominator of which is also an integer and a fraction, etc. Its general form may be put as follows:

$$a+\cfrac{b}{c+\cfrac{d}{e+\cfrac{f}{g+\text{ etc.}}}}$$

Which is also written in more convenient form as

$$a+\frac{b}{c}+\frac{c}{d}+\frac{d}{e}+\frac{e}{f}+\ldots\ldots$$

The number of terms may be finite (terminating) or infinite (non-terminating). The transcedental numbers like π, can be expressed in the continued fraction form.

Continued Product. Refers to the product of a finite or infinite number of terms, represented as $\prod_{n=1}^{m} U_n$ or $\prod_{n=1}^{m} U_n$ respectively. The transcedental number π can be expressed in the form of a continued product having infinite number of terms.

Continuous Function. A function f(x) is called continuous at a point x = a if $\lim_{x\to a} f(x)$ exists and is equal to f(a). If this is not satisfied, then the function is discontinuous at x = a, or has discontinuity there. f(x) is considered to be continuous in the interval from x = a to x = b if it is continuous at all the points of this interval. For example, sin θ, cos θ are continuous at every point while tan θ is discontinuous

at $\theta = \pm\frac{\pi}{2}, \pm\frac{3\pi}{2}$, etc.

Continuum. Suppose a line is drawn infinitely on both side and having an origin. Now every point on the line corresponds to some real number and to every number three corresponds some point on the line. The points on the line are known as continuum of one dimension. Similarly, time is considered as a continuum.

Convergent Sequence. Refers to a sequence $\{x_n\}$ which is said to be convergent to a quantity x if $\lim_{n\to\infty} (x_n) = x$.

For example, the sequence $\left\{ 1, \frac{1}{2}, \frac{1}{4}, \frac{1}{8}, \ldots \right\}$ converges to zero.

Converse. Means a logical implication taken in the reverse order. For example, the converse of

if I am under 10 and I go to school is

if I go to school, then I am under 10. The converse of an implication is not always true even if the implication itself is true. There are a number of theorems in mathematics for which both the statement and the converse are true. For example, the theorem:

if two chords of a circle are equidistant from the centre then they are equal has a true converse :

if two chords of a circle are equal, then they are equidistant from the centre of the circle.

Conversion Graph. Refers to a graph that depict a relationship between two variable quantities. If one quantity is known, the corresponding value of the other can be read directly from the graph. For example, air pressure depends on height above sea level. A standard curve of altitude against air pressure can be plotted on a graph. An air-pressure measurement can be converted to an indication of height by reading the appropriate value from the graph.

Co-ordinate. Refers to one of a set of numbers which locate a point in space. If the point is known to be on a given line, only one coordinate is needed : if in a plane, two are required; if in space, three coordinates are required.

Coordinate Paper. Refers to paper ruled with graduated rulings in order to help in plotting points and drawing the loci of equations.

Coordinate Systems. Geometrical system that locate points in space by a set of numbers.

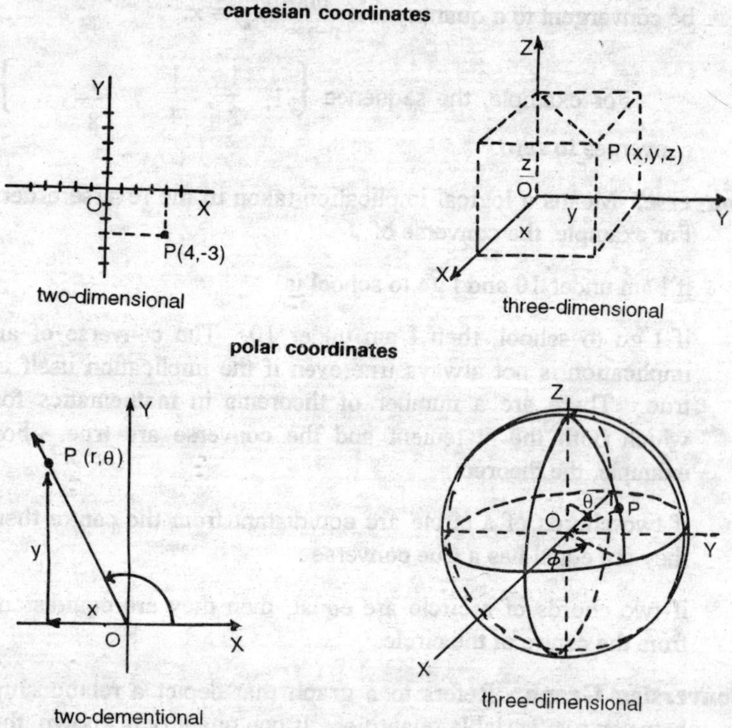

Fig. C-4.

Coordinate systems. In two-dimensional Cartesian coordinates point P is located by giving its x- and y-coordinates, in this case 4 and —3(always given in this order, as shown). In three dimensional Cartesian coordinates P is located in terms of three axes. In polar coordinates P is located by a radius r and an angle θ. In three dimensions a second angle ϕ is required. In the diagrams Cartesian coordinates are superimposed on the polar coordinates.

In **Cartesian coordinates,** devised by Rene *Descartes, a point is located by its distance from intersecting lines called axes. In a plane (two dimensions) there are two axes and in space (three dimensions)there are three. Usually the axes are at right angles to each other and are known as rectangular axes, but oblique axes are also used in exceptional circumstances.

Polar coordinates denote position by distance and direction. A fixed point, called the origin, and a fixed line, called the polar axis, are taken as the references. For any point, the polar coordinates are the length, r, of the radius of the circle centred at the origin and passing through the point, and the angle, θ, between this radius and the polar axis. In three dimensions, spherical polar coordinates are used. The radius of a sphere centred at the origin and the two angles it makes with the polar axis define the point.

Coplanar. Means lying in the same plane. e.g., *coplanar lines* are lines which lie in the same plane; *coplanar points* are points which lie in the same plane. Three points are necessarily coplanar; four points described by their rectangular Cartesian coordinates are coplanar if and only if the following determinant is zero (otherwise, the absolute value of the determinant is the volume of a parallelopiped with the four points as four of the eight vertices and three of the points adjacent to the fourth):

$$\begin{vmatrix} x_1 & y_1 & z_1 & 1 \\ x_2 & y_2 & z_2 & 1 \\ x_3 & y_3 & z_3 & 1 \\ x_4 & y_4 & z_4 & 1 \end{vmatrix}$$

Cornu-Hartman Formula. A relation between the deviation D produced by a prism and the wavelength λ of the light passing through it.

$$\lambda = \lambda_0 + \frac{C}{C - D_0}$$

where λ_0, C and D_0 are constants.

Cornu's Spiral. A double spiral used in computing intensities in a Fresnel diffraction pattern, obtained by plotting

$$x = \int_0^v \cos \frac{\pi v^2}{2} dv \text{ against } y = \int_0^v \sin \frac{\pi v^2}{2} dv$$

Corollary. Refers to a throrem that follows so obviously from the proof of some other theorem that no, proof is necessary, a byproduct of another theorem.

Correct. Means without error in principle or computation. One speaks of a correct proof, correct solution, correct answer or correct computation.

Correlation. In pure mathematics. A linear transformation which, in the plane, carries points into lines and lines into points and, in space, carries points into planes and planes into points.

Correspondence. Means *Mapping or Function.* Refers to relationship between some or all elements of one set and some or all elements of another set.

Cosine Formula. In any triangle ABC if a, b, c are the length of its sides we have

$$\cos A = \frac{b^2 + c^2 - a^2}{2bc}, \cos B = \frac{c^2 + a^2 - b^2}{2ca},$$

$$\cos C = \frac{a^2 + b^2 - c^2}{2ab}.$$

Cotes' Rule. For obtaining the area under a curve, assume for y

$$y = A_0 + A_1 x + A_2 x^2 + \ldots + A_{n-1} x^{n-1}$$

Determine the coefficients $A_1, A_2, \ldots A_{n-1}$ so that, for the n equidistant values of x, y shall have the prescribed values y_1, y_2, ... y_n. The area is then given by

$$\int y\, dx = A_0 x + \frac{1}{2} A_1 x^2 + \frac{1}{3} A_2 x^3 + \ldots + \frac{1}{n} A_{n-1} x^n$$

taken betwen proper limits of x.

Coulomb Symbol : C. The SI unit of electric charge. It is equal to the charge carried by an electric current of one ampere flowing for one second. 1 C = 1 As.

Countable Set. Denumerable Set. Refers to any set whose elements have one-one correspondence with the elements of the set of positive integers.

Couple. Refers to two parallel forces which are equal in magnitude and opposite in direction, they form a couple. The couple when acted upon a body makes the body to turn about an axis perpendicular to the plane of forces. If F represents the magnitude of each of the forces and d the arm of the couple then the *moment* of the couple (*torque*) about any point in the plane is given *Fd.*

Covariance. A measure of association between two variables. If x and y are any two variables with x and y their means, n is the number of pairs of values of x and y, then the covariance between the variables is given by

$$\Sigma\, [\overline{(x_i - x)\,(y_i - \overline{y}\,)}]/n.$$

Crammer's Rule. Refers to a rule that is used for solving simultaneous equations by determinant method. For example, if we have three simultaneous algebraic equations.

$$a_1x + b_1y + c_1z = d_1,\ a_2x + b_2y + c_2z = d_2,\ a_3x + b_3y + c_3z = d_3.$$

$$\text{and if } D = \begin{vmatrix} a_1 & b_1 & c_1 \\ a_2 & b_2 & c_2 \\ a_3 & b_3 & c_3 \end{vmatrix} \neq 0,\ D_1 = \begin{vmatrix} b_1 & c_1 & d_1 \\ b_2 & c_2 & d_2 \\ b_3 & c_3 & d_3 \end{vmatrix}$$

$$D_2 = \begin{vmatrix} a_1 & c_1 & d_1 \\ a_2 & c_2 & d_2 \\ a_3 & c_3 & d_3 \end{vmatrix},\quad D_3 = \begin{vmatrix} a_1 & b_1 & d_1 \\ a_2 & b_2 & d_2 \\ a_3 & b_3 & d_3 \end{vmatrix}$$

then $x = \dfrac{D_1}{D}$, $y = \dfrac{-D_2}{D}$, $z = \dfrac{D_3}{D}$ $(D \neq 0)$. This is called Crammer's rule. It can also be used in a general case of n algebraic equations having n unknowns.

Critical Point. Refers to any point on the curve $y = f(x)$ where the gradient $\frac{dy}{dx}$ is either zero or infinite i.e., where the tangent to the curve is parallel to the either of the coordinate axes.

Cross Multiplication. Means replacing the equality of the ratios, i.e., $\frac{a}{b} = \frac{c}{d}$ by the equivalent statement

$$a.\ d = b.\ c.$$

Cubic Equation. Refers to an algebraic equation of third degree in the form $a_0x^3 + a_1x^2 + a_2x + a_3 = 0$ where $a_0 \neq 0$. It has three roots out of these at least one root is real. If only one root is real the other two will be complex roots conjugate of each other. The roots of the equation $x^3 = 1$ and $1.\ \omega$ and ω^2 where $1 + \omega + \omega = 0$.

Cuboid. Refers to a box shaped solid figure bounded by six rectangular faces. The opposite faces are congruent and parallel. At each of the eight vertices, three faces meet at right-angles to each other. The volume of a cuboid is given by length × breadth × height. The surface area is the sum of the areas of the faces, that is

$$2\,(l \times b) + 2\,(b \times h) + 2\,(l \times h).$$

In the special case in which $l = b = h$, all the faces are square and the cuboid is a cube having volume l^3 and surface area l^2.

Cumulative Frequency. Refers to the total frequency of all values up to and including the upper boundary of the class interval that is under consideration.

Curl (Symbol ∇x). Refers to a vector operator on a vector function that, for a three-dimensional function, is equal to the sum of the vector produce (cross product) of the unit vectors and partial derivatives in each of the component directions. That is

$$\text{curl } F = \nabla x F = i \times \partial F/\partial x + j \times \partial F/\partial y + k \times \partial F/\partial z$$

where i, j and k are the unit vectors in the x, y and z directions respectively. In physics, the curl of a vector arises in the relationship between electric current and magnetic flux, and in the relationship between the velocity and angular momentum of a moving fluid.

Curvature. Refers to the rate of change of the slope of the tangent to a curve, as compared to distance along the curve. For each point on a smooth curve there is a circle which has the same tangent and the same curvature at that point. The radius of this circle, known as the ***radius of curvature***, is the reciprocal of the curvature, and its centre is called the ***centre of curvature***. If the graph of a function $y = f(x)$ is a continuous curve then the slope of the tangent at any point it given by *dy/dx* and the curvature is given by

$$d/dx\,[\tan^{-1} dy/dx].$$

Cusp. Refer to a double point where there are more than one tangents that are real and coincident. Cusps are of following types: (a) single cusp of first species, (b)single cusp of second species, (c) double cusp of first species, (d) double cusp of second species, (e) double cusp of both species.

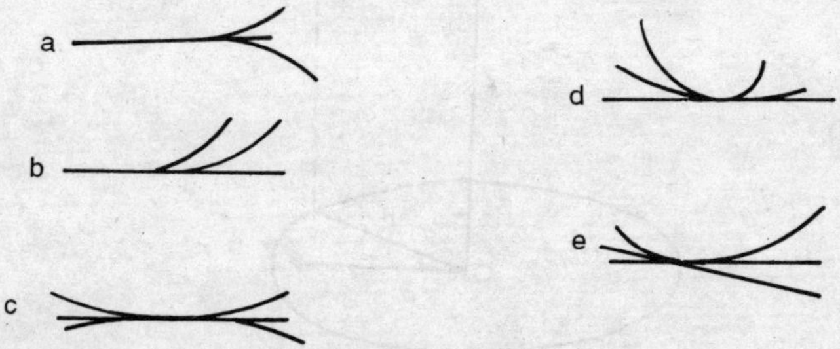

Fig. C-5.

Cyclic Polygon. Refer to a polygon having all its vertices on the circumference of a circle. All triangles are cyclic, squares and rectangles are cyclic quadrilaterals, all regular polygons are cyclic.

Cycloid. The curve traced out by a point on the circumference of a circle as it rolls along a flat surface.

Cylinder. Refers to a solid that is defined by a closed curve known as *guiding curve* (forming a base) or *directrix*, having an identical curve parallel to it. Any line segment from a point on one curve to a corresponding point on the other curve is known as an *element* of the cylinder. If one such element moves parallel to itself round the base it generates the curved lateral surface. The line is known as generator of the cylinder. If the base is a circle then the cylinder is a circular one. The line joining the centres of such circular bases is known as *axis* of the circular cylinder. If the axis is perpendicular to the circular base the cylinder is known as the right circular cylinder. The volume of a right circular cylinder having radius of its base as r and its height h is $\pi r^2 h$ and the curved lateral surface is $2\pi rh$.

Cylindrical Coordinates. Refers to coordinates of a point in space as (r, θ, z) where r is its horizontal distance from fixed vertical

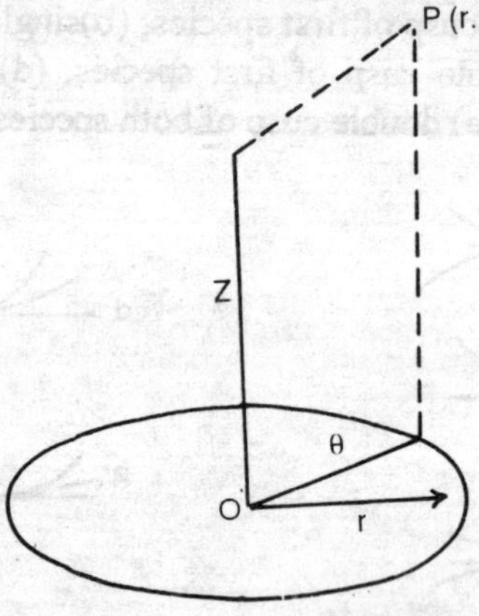

Fig. C-6.

axis known z-axis, θ is the angle, the horizontal radius made with an axis in the horizontal plane known as X-axis and z is the height of the point above the fixed horizontal plane known as X — Y plane.

For a point P(r, θ, z) the corresponding rectangular cartesian coordinates (x, y, z) are $x = r \cos \theta$, $y = r \sin \theta$ and $z = z$.

The coordinates are described as cylindrical since the point describes in cylindrical surface when r is constant.

D

D'Alembertian. A differential operator

$$\Box^2 V = \frac{\partial^2 V}{\partial x^2} + \frac{\partial^2 V}{\partial y^2} + \frac{\partial^2 V}{\partial z^2} - \frac{1}{c^2}\frac{\partial^2 V}{\partial t^2}$$

Thus $\Box^2 V = 0$ is the wave equation for waves travelling with a velocity c.

D'Alembert's Principle. For a system of particles of mass m_i, acted upon by the real forces F_i subject to constraints such that, at a certain instant, they have acceleration a_i, then

$$F_i - ma_i = 0 \text{ for all } i.$$

ma_i is sometimes called the 'effective force' acting on the particle, in which case D'Alembert's Principle can be stated as: the reversed effective forces and the real forces together give statical equilibrium.

D'Alembert's Ratio Test (Generalized Ratio Test). A method of showing if a series is convergent or divergent. The absolute value of the ratio of each term to the one before it is taken:

$$| U_{n+1} / U_n |$$

If the limit of this is l as n tends to infinity and l is less than 1, then the series is convergent. If *l* is greater than 1, the series is divergent. If *l* is equal to 1, this test fails and some other method has to be used.

D'Alembert's Test for Convergence

In $\qquad \Sigma u_n = u_1 + u_2 + \ldots$

if for all values of n greater than r

$$\left| \frac{u_n + 1}{u_n} \right| < \rho$$

where ρ is a positive number less than unity and independent of n, Σu_n is absolutely convergent.

Damping. Refers to the reduction in amplitude of a vibration with time due to some form of resistance. A swinging pendulum will finally come to rest; a plucked string will not vibrate for long-in both these cases internal and/or external resistive forces progressively reduce the amplitude and bring the system to equilibrium.

In many cases the damping force(s) are proportional to the object's speed. In any event, energy must be transferred from the vibrating system so as to overcome the resistance. Where damping is an asset (as in bringing the pointer of a measuring instrument to rest), the optimum situation occurs when the motion comes to zero in the shortest possible time without vibration. This is critical *damping*. If the resistive force is such that the time taken is longer than this, *overdamping* takes place. Conversely, *underdamping* involves a longer time having vibrations of decreasing amplitude.

Data Now often used as a singular noun. The facts that refer to or describe an object, idea, condition, situation, etc. In computing, data can be considered as the facts on which a program operates as opposed to the instructions in the program. It can only be accepted and processed by the computer in binary form. Data is sometimes considered to be numerical information only.

Database. A collection of data held on computer so that it can be accessed in different ways. For example, all the entries in this encyclopaedia are held on a database and it is possible for the editors to produce lists of types of entry; for example, all people who are still alive, or all US presidents, or all chemical compounds. These are three types of database classified according to the way the records are structrured: *hierarchical, network, or relational*. The software used for accessing, changing, and deleting data is a *database management system* (DBMS).

Data Processing. The organization, transmision and storage of information. **Automatic data processing** usually refers to systems using punch-card machines, paper tape, magnetic tape, etc., as opposed to **electronic data processing,** which is based on electronic computers. In batch processing, data is grouped and coded before processing. The alternative is on-line processing in which each user feeds data into the system continuously. Many system incorporate both of these forms of processing for different types of works.

Data Protection. Legislation to protect individuals who have information about them stored on computers. The Data Protection Convention, signed by members of the EC in 1981, led to the British Data Protection Act (1986). This obliges all computer users who are storing personal information to register with a central body, and to maintain certain standards of security and confidentiality. Individuals have the right to see their entries and, under certain circumstances, to ask for amendments or deletion.

Deci. A prefix standing for 10^{-1}. For example,

$$1 \text{ decimetre} = \frac{1}{10} \text{metre.}$$

Decibel. Symbol : dB. A unit of power level, generally of sound wave or electrical signal, measured on a logarithmic scale. The threshold of hearing is taken as 0 dB in sound measurement.

Ten times this power level is 10 dB. The fundamental unit is the bel, but the decibel is almost exclusively used (1db = 0.1 bel)

A power P has a power level in decibels given by :

$$10 \log_{10} (P/ P_0)$$

where P_0 is the reference power.

Decimal. Refers to any numeral in the *decimal number system* but sometimes restricted to a **decimal fraction** (i.e., a number that in decimal notation has no digits other than zeros to the left of the decimal point. **A mixed decimal** is a decimal plus an integer, as 23.35. Decimals having the same number decimal places, as 2.361 and .253, are **similar.** Any decimals can be made similar by annexing the proper number of zeros; e.g., 36 can be made similar to .321 by writing it as .360.

Decimal Expansion. For a real number, the representation of the number making use of the decimal number system.

Decimal Measure. Any system of measuring wherein each unit is derived from some standard unit by multiplying or dividing the latter by some power of 10.

Decimal Place. Refers to the position of a digit in a decimal e.g., in 123.456, 1 is in the *hundred's place,* 2 is in the *ten's place,* 3 is in the *unit's place* 4, is in the *first decimal place,* 5 is in the *second decimal place,* and 6 is in the *third decimal place.*

Decimal System. The number system in common use, having a base 10 and then thus using ten separate numerals. It also involves the use of a decimal point to express numbers less than one, instead of the method of fractions; for example it is expressed as 0.25. The decimal system was invented by the Hindus and adopted by the Arabs in the 9th century. The use of decimal fractions originated in Italy in the 12th century but was first formalized by the mathematician Simon Stevin (1548-1620) in 1585. The use of the decimal point did not occur until the beginning of the 18th century. Decimalization of currency systems was introduced by France after the Revolution and followed by most other European and American countries, except for Britain, which did not decimalize until 1971.

Decreasing Function of one Variable. Refers to a function whose value decreases as the independent variable increases; a function whose graph falls as the abscissa increases. If function is differentiable on an interval I, then the function is decreasing on I if the derivative is non-positive throughout I and is not identically zero in any interval. A decreasing function is said to be **strictly decreasing** to distinguish it from a *monotone decreasing* function. A function f is **strictly decreasing** on an interval (a, b) if

$$f(y) < f(x)$$

for all numbers x and y of I with $x < y$; f is **monotone decreasing** on I if $f(y) \leq f(x)$ for all numbers x and y of I with $x < y$.

Dedekind's Definition. Suppose a certain rule devides a whole system of rationals into two classes, a lower class A and an Upper class A', so that any number in A is less than any number of A', Either

(1) A has greatest number or A' has a least number, in which case the classification defines an irrational number which is to follow all in A or the least in A', or

(2) if there is no greatest number in A and no least number in A', the classification defines an irrational number is to flollow all the numbers in A and to precede all in A'.

Deduction. Refers to a series of logical steps in which a conclusion is drawn directly from a set of initial statement (premisses). A deduction is valid if a sentence or statement that asserts the premisses and denies the conclusion is a contradiction.

Definite Integral (Riemann integral). The result of integrating any function of a single variable, f(x), between two specified values of x : x_1 and x_2. The definite integral of f(x) is written

$$\int_{x_1}^{x_2} f(x)\,dx$$

If the general expression for the integral of f(x) (its indefinite integral) is another function of x, g (x), the definite integral is given by :

$$g(x_1) - g(x_2)$$

Degenerate Conic. Refers to the conic sections which are neither elliptic, parabolic, nor hyperbolic.

Degree. (1) A unit of plane angle equal to 1/360th of a complete revolution. It is subdivided into 60 minutes, each of which consists of 60 seconds. (2) An interval on a temperature scale 1° on the Celsius (centigrade) scale is equal to one-hundredth of the difference in temperature between freezing and boiling water. It is also equal in magnitude to 1 Kelvin 1° on the Fahrenheit scale is 9/5 times the Celsius degree.

De Gua's Rule. If a group of r consecutive terms is missing from a polynomial f(x) = 0, then

(1) if r is even, the equation has at least r imaginary roots;

(2) if r is odd, they are at least r + 1 or at least r — 1 imaginary roots according as the terms which immediately precede and follow the group have like or unlike signs.

De 1-Hopital's Rule. The rule stating that the limit of the ratio of two functions of the same variable (x) as x approaches a value a, is equal to the limit of the ratios of their derivatives with respect to x. That is, the limit of f(x)/g(dx) as $x \to a$ is the limit of f' (x)/g' (fx) as $x \to a$.

This rule can be used to find the limit of f (x) / g(x) at points where both f(x) and g(x) are zero and the ratio is thus indeterminate. Any function which gives rise to an indeterminate form, and which be expressed as a ratio of two functions, can be dealt with in this way. For example, in

writing $F(x) = (x^2 - 3) / (x - 3)$

$f(x) = (x^2 - 3)$

and $g(x) = (x - 3)$

gives $F(x) = f(x) / g(x)$.

The limit of F(x) as $x \to 3$ is indeterminate (since $(x - 3) = 0$). It can be found by using the limit of

$$f'(x) / g'(x) = 2x$$

as $x \to 3$. Thus the limit is 6.

If f' (x) /g' (x) also gives an indeterminate form at x = a, this rule can be used again, differentiating as many times as is necessary.

Delambre's Analogies (or Gauss' Analogies). In the solution of spherical triangles, if a, b and c are the sides of the triangle and α, β and γ are the angles opposite a, b and c respectively, then

$$\sin \tfrac{1}{2}\alpha \cos \tfrac{1}{2}(b + c) = \cos \tfrac{1}{2}a \cos \tfrac{1}{2}(\beta + \gamma)$$

$$\cos \tfrac{1}{2}\alpha \sin \tfrac{1}{2}(b - c) = \sin \tfrac{1}{2}a \sin \tfrac{1}{2}(\beta - \gamma)$$

$$\sin \tfrac{1}{2}\alpha \sin \tfrac{1}{2}(b + c) = \sin \tfrac{1}{2}a \cos \tfrac{1}{2}(\beta - \gamma)$$

$$\cos \tfrac{1}{2}\alpha \cos \tfrac{1}{2}(b - c) = \cos \tfrac{1}{2}a \sin \tfrac{1}{2}(\beta + \gamma)$$

Also **Napier's Analogies :**

$$\tan \tfrac{1}{2}(b+c)\cos \tfrac{1}{2}(\beta+\gamma) = \tan \tfrac{1}{2} a \cos \tfrac{1}{2}(\beta-\gamma)$$

$$\tan \tfrac{1}{2}(\beta+\gamma)\cos \tfrac{1}{2}(b+c) = \cot \tfrac{1}{2} \alpha \cos \tfrac{1}{2}(b-c)$$

$$\tan \tfrac{1}{2}(b-c)\sin \tfrac{1}{2}(\beta+\gamma) = \tan \tfrac{1}{2} a \sin \tfrac{1}{2}(\beta-\gamma)$$

$$\tan \tfrac{1}{2}(\beta-\gamma)\sin \tfrac{1}{2}(b+c) = \cot \tfrac{1}{2} \alpha \sin \tfrac{1}{2}(b-c).$$

De-Moivres' Theorem. According to this theorem for all n, positive or negative, integer or fraction the value or one of the value or $(\cos\theta + i\sin\theta)^n$ is $\cos n\theta + i\sin\theta$. It is used calculating the value of a complex number raised to some power.

De Morgan's Laws. If A, B, C, ...are subsets of a set R and a dash represents the complement of a set, then

$$(A \cup B \cup C \,....)' = A' \cap B' \cap C' \cap \,......$$

$$(A \cap B \cap C \,.....)' = A' \cup B' \cup C' \cup \,....$$

De Morgan's Rules. These apply to **Boolean Algebra.** Any logical or binary expression equals the complement(or negation) of the expression obtained by changing all logical products (sometimes called AND statements) to logical sums (sometimes called OR statements) or vice versa and replacing variables with their complements.ie.

$$\overline{A+B} = \overline{A}\,.\,\overline{B} \text{ and } \overline{A.\,B} = \overline{A} + \overline{B}$$

where $\overline{A}$ and $\overline{B}$ are the complements of A and B respectively.

De Morgan's Test. A series Σu_n in which $\lim_{n\to\infty}\left|\frac{u_{n+1}}{u_n}\right| = 1$ will be absolutely convergent if a positive number c exists such that

$$\lim_{n\to\infty}\left\{\left|\frac{u_{n+1}}{u_n}\right| - 1\right\} = -1 - c.$$

Denumerable Set. Refers to a set wherein the elements can be counted. For example, the set of prime numbers, although infinite, can be counted, as can the set of positive even integers. These are called as denumerably infinite sets. The set of rational numbers, on the other hand, is not denumerable as between any two elements there can always be found a third.

Dense Set. A set E in a space M is **dense** (or **dense in M,** or **everywhere dense**) if every point of M is a point of E or a limit point of E, or (equivalently) if the closure of E is M, or if every neighbourhood in M contains a point of E. A set E is **dense in itself** if every point of E is an accumulation point of E; i.e., if each neighbourhood of any point of E contains another point of E. A set E is **nondense** (or **nowhere dense**) relative to M if no neighbourhood in M is contained in the closure of E, or (equivalently) if the complement of the closure of E is dense in M. The set of rational number is dense in itself and dense in the set R of all real numbers, as is also the set of irrational numbers. This is equivalent to the fact that between any two real numbers (either rational or irrational) there are both rational numbers and irrational numbers. However, the set S = { 0, 1, 1/2, 1/4, 3/4,... } is nowhere dense in R, since 0 is the only accumulation point of S and therefore no neighbourhood in R contains only accumulation points of S.

Dependence, Domain of Dependence. For an initial-value problem for a partial differential equation, the value of the solution at a point p and time *t* can be determined by the initial values on only a portion of their entire range, known as the *domain of dependence* e.g., for the wave equation $(1/c^2)\, u_u = u_{xx}$,with initial conditions u(x, 0) = f(x), ut (x, 0) = g(x) ,the value of the solution at the point x and time t depends only on the initial values in the interval [x — bt, x + ct].

Dependent, Dependent Equation. An equation is dependent on a set of equations if it is satisfied by every set of values of the unknowns that satisfy all the other equations. One of three linear equations in two unknowns is dependent on the other two if the graphs of these two are not coincident and the three graphs are concurrent lines. A set of equations is **dependent** if some member of the set is dependent on the others. When solving a system of equaions, any dependent equation can be discarded.

Dependent Functions. A set of functions one of which can be expressed as a function of the others. e.g., v = sin u and thus { u, v} is a set of dependent functions if

$$u(x, y) = \frac{x+1}{y+1}, \quad v(x, y) = \sin \frac{x+1}{y+1}.$$

Also, f_1 and f_2 are dependent if $f_1(x) = x^2$ and $f_2(x) = x^2$ for all x, since $f_1 = f_2$.

Linearly Dependent. A set of objects $z_1, z_2, \ldots z_n$ (vectors, matrices, polynomials, etc.) is **linearly dependent** on a given set if there is a linear combination of them,

$$a_1 z_1 + a_2 z_2 + \ldots + a_n z_n,$$

which is zero (with the coefficients in the given set), but at least one of the coefficient is non-zero. The dependence is relative to the nature of the coefficients allowed. A set of objects is **linearly independent** if it is not linearly dependent. The binomials x + 2y and 3x + 6y are linearly dependent, since

$$-3(x + 2y) + (3x - 6y) = 0.$$

Depreciation Charge. Refers to a decrease in the book value, generally annual, such that the total of these decreases, without intertest, will equal the original book value (or cost) minus the scrap value at the end of a certain number of years (the **total depreciation**). There are different methods for calculating depreciation charges. For the **straight line method,** equal depreciation charges are made each year. For the **declining balancing method** (or **constant percentage method**) each depreciation charge is calculated as a constant percent of the book value at the time of computation; this percent is equal to

$$(1 - \sqrt[n]{R/C}) \times 100,$$

where C is the cost, R the scrap value, and n the number of years.

De-Pressed', Depressed Equation. Refers to the equation obtained by reducing the number of roots in an equation, i.e., by dividing out the difference of the unknown and a root; e.g., $x^2 - 2x + 2 = 0$ is the *depressed* equation obtained from $x^3 - 3x^2 + 4x - 2 = 0$ by dividing the left member of the latter by $x - 1$.

Derivative, Differential Coefficient. *Result of differentiation*. If y is a function of x, given by y = f(x), the first order derivative of y is $\frac{dy}{dx}[= f'(x)]$. If y = u.v. where u and v are functions of x then

$$\frac{dy}{dx} = \frac{du}{dx}.v + u.\frac{dv}{dx}$$

$$\text{and if } y = \frac{u}{v} \text{ then } \frac{dy}{dx} = \left(v\,\frac{du}{dx} = u\,\frac{dv}{dx} \right) / v^2.$$

Higher order derivatives can also be defined, so that second order derivative is $\frac{d}{dx}\left(\frac{dy}{dx}\right)$ represented by $\frac{d^2y}{dx^2}$ or by y^n.

The nth order derivative of y is denoted by $\frac{d^ny}{dx^n}$ or by y^n.

Descartes' Rule of Signs. Refers to a rule that is used for determining an upper bound to the number of positive zeros and to the number of negative zeros of a polynomial. The rule states that the number of positive zeros of a polynomial p(x) either is equal to the number of its variations of sign or is less than that number by an even integer, a zero of multiplicity m being counted as m zeros. The polynomial can be tested for negative zeros by applying this rule to the polynomial p(— x). For example, the polynomial

$$x^4 - x^3 - x^2 + x - 1$$

has three variations in sign and hence has either three or one positive zeros. Since replacing x by — x gives $x^2 + x^3 + x^2 - x - 1$, which has only one variation of sign, the original polynomial has exactly one negative zero.

De-Tachment, Rule of Detachment. If an implication is true and the antecedent is true, then the consequent is true, e.g., if the statements " If may team lost, I will eat my hat" and "My team lost" are both true, then the statement "I will eat my hat" is also true.

Determinant. Refers to a square array of quantities called *elements.* If there are n^2 elements, there must be n rows and n columns, and n is known as the *order of the determinant.* For example, the second order determinant, is

$$\begin{vmatrix} a_1 & b_1 \\ a_2 & b_2 \end{vmatrix} = a_1 b_2 - a_2 b_1.$$

Third order determinant is

$$\begin{vmatrix} a_1 & b_1 & c_1 \\ a_2 & b_2 & c_2 \\ a_3 & b_3 & c_3 \end{vmatrix}$$

$$= a_1 b_2 c_3 - a_1 b_3 c_1 + a_3 c_2 b_1 - b_1 a_2 c_3 + c_1 a_2 b_3 - c_1 a_3 b_2$$

$$= a_1 \begin{vmatrix} b_2 & c_2 \\ b_3 & c_3 \end{vmatrix} - a_2 \begin{vmatrix} b_1 & c_1 \\ b_3 & c_3 \end{vmatrix} + a_3 \begin{vmatrix} b_1 & c_1 \\ b_2 & c_2 \end{vmatrix}$$

$$= a_1 A_1 - a_2 A_2 + a_3 A_3$$

where

$$A_1 = \begin{vmatrix} b_2 & c_2 \\ b_3 & c_3 \end{vmatrix} \quad A_2 = \begin{vmatrix} b_1 & c_1 \\ b_3 & c_3 \end{vmatrix} \quad A_3 = \begin{vmatrix} b_1 & c_1 \\ b_2 & c_2 \end{vmatrix}$$

where A_1, A_2, A_3 are *minors* of the determinant.

Determinants are expanded by the *rule of Sarrus.*

Cramer's Rule uses these notations for solving simultaneous equations.

Mean Deviation. The *mean* or *expected value* of $| X - \mu |$ where μ is the mean. For a continuous random variable having probability densitgy function f, the **mean deviation** is given by

$$\int_{-\infty}^{\infty} | x - \mu | f(x)\, dx,$$

if this integral converges. For a discrete random variable with values $\{x_n\}$ and probability function p, the **mean deviation** is given by $\Sigma \mid x_i - \mu \mid p(x_i)$, if this sum converges; sometimes the median is used for μ. If a random variable has a finite number of values having equal probabilities, then the sum of deviations, with regard to sign, is zero about the mean and, without regard to sign, it is minimal about the median.

Mean-Square Deviation. The second moment about some number a. If a is the mean, the mean-square deviation is the *variance.*

Probable Deviation. Refers to the product of approximately 0.6745 and the *standard error;* for a normal random variable, the probability of a deviation having absolute value greater than this is 1/2. No longer used as being replaced by the standard error.

Quartile Deviation. Refers to one-half of the difference between the upper and lower quartiles : $1/2\,(Q_3 - Q_1)$.

Standard Deviation. For a random variable (or the associated distribution function,) the positive square root of the *variance* (*see* variance).

d'Huilier's Equation. In the solution of spherical triangles

$$\tan\left(\frac{\alpha}{2}-\frac{\varepsilon}{4}\right) = +\left[\frac{\tan \frac{1}{2}(s-b)\ \tan \frac{1}{2}(s-c)}{\tan \frac{1}{2} s\ \tan \frac{1}{2}(s-a)}\right]$$

where ε is the spherical excess ($=\alpha+\beta+\gamma-\pi$) and $s = \frac{1}{2}(a+b+c)$.

Differential. An infinitesimal change in an independent variable or in a dependent variable due to a small change in independent variable, dx and dy are differentials in $\frac{dy}{dx}$, the differential coefficient of y w.r.t.x.

Differential Coefficient. *Derivative.* For a function $y = f(x)$ the differential coefficient of y w.r.t. x is defined as

$$\frac{dy}{dx} = \lim_{\delta x \to \infty} \frac{f(x+\delta x) - f(x)}{\delta x} = \lim_{\delta x \to \infty} \frac{\delta y}{dx}.$$

It is also represented by y' or Dy where $D = \frac{d}{dx}$.

Differential Equation. Refers to any equation that contains derivatives. For example, $dy/dx + 4y + 6x - 17 = 0$, is a differential equation. The order of the highest order derivative is known as the order of the *differential equation* and the degree of the differential equation is the degree of the highest order derivative. The above differential equation is of first order and of first degree.

Differential Geometry. Refers to the branch of mathematics in which calculus is used to the theory of waves and surfaces in a three demensional space.

Defferentiation. Refers to the process of finding the derivative of function. It stands for finding the rate of change of one variable with respect to another. For example, if a particle has traversed a distances s at any time t and a distances $s + \delta s$ at any time $t + \delta t$, the velocity of the particle at a time is given by $\frac{ds}{dt} = \lim_{\delta t \to \infty} = \frac{\delta s}{\delta t}$. Similarly $\frac{d^2s}{dt^2}$ gives the acceleration of the particle at any time t, that is the rate of change of velocity.

Digital. Using numerical digits. For example, a digital watch shows the time in numbers of hours and minutes and not as the position of hands on a dial. In general, digital devices work by some kind of counting process, either mechanical or electronic. The abacus is a very simple example. Early calculating machines counted with mechanical relays. Modern calculators use electronic switching circuits.

Dihedral. Refers to a figure produced by two intertsecting planes. The straight line wherein two planes intert set is known as the *edge* of dihedral and the two planes are known as its *faces*. The dihedral angle is the angle between its two faces, which is measured by the angle between the line of cross-section.

Dimension. Refers to (1) the least number of coordinates required to represent a point in a plane or in space. A point in a plane needs two coordinates hence a plane figure will be two dimensional. Similarly, a solid is three dimensional. Some abstract spaces are considered as n-dimensional. (2) The size of a plane figure or a solid. For instance the dimension of a circle is its radius, that of a rectangle are its length and breadth, and of a rectangular parallelopiped are its length, breadth and height. (3) If L is the dimension of length, M that of mass and T that of time then the dimension of volume is $[L]^3$, dimension of velocity is $[L]\ [T]^{-1}$ that of force is $[M]\ [L]\ [T]^{-2}$ that of stress is $[M]\ [L]^{-1}\ [T]^{-2}$. (4) Dimensions of a matrix are its number or rows and those of its columns.

Dimensional Analysis. Refers to the use of the dimensions of physical quantities to check relationships between them. For example, Einstein's equation $E = mc^2$ can be checked. The dimensions of speed are $([L]\ [T]^{-1})^2$; i.e., $[L]^2\ [T]^{-2}$, so mc^2 has dimensions of $[M]\ [L]^3\ [T]^{-2}$. Energy has these dimensions because it is force $[M]\ [L]\ [T]^{-2}$ multiplied by distance $[L]$. Dimensional analysis is also used to get the units of a quantity and to suggest new equations.

Diocles, Cissoid of. A curve whose equation is

$$y^2\,(a - x) = x^3$$

or alternatively

$$r = a\left(\frac{1}{\cos\phi} - \cos\phi\right)$$

Directed. Means having a specified positive or negative sign, or a definite direction. A directed number generally has one of the signs + or — written in front of it. A directed angle is measured from one specified line to the other. If the direction were reversed, the size of the angle would be a negative number.

Direction Cosines. Refers to *Cosines of the direction angles.* If a line makes angles α, β and γ with the x-axis, y-axis, z-axis respectively then direction cosines of the line are $\cos\alpha$, $\cos\beta$, $\cos\gamma$. These are generally represented by letters l, m, n, so that $l = \cos\alpha$, $m = \cos\beta$, $n = \cos\gamma$.

Directional Derivative. Refers to rate of change of a function with respect to distance s in a specified direction or along specified curve. If α, β, γ are the angle formed by the specified direction or the tangent to the specified curve at the pointp (x,y,z) having the three coordinate axes then the directional derivative of a function f(x, y, z) is

$$\frac{df}{ds} = \frac{\partial f}{\partial x}\cos\alpha + \frac{\partial f}{\partial y}\cos\beta + \frac{\partial f}{\partial z}\cos\gamma .$$

The directional derivative is a maximum in direction of the normal to the surface and is known as the gradient of f represented by ∇f.

Director Circle. The locus of the point oif intersection of a pair of tangents to a centreal conic which are at right angles to each other. For the central conic.

$$ax^2 + 2hxy + by^2 + 2gx + 2fy + c = 0$$

the equation of the director circle is

$$(ab - h^2)(x^2 + y^2) - 2(hf - bg)x - 2(gh - af)y + c(a + b) - (f^2 + g^2) = 0.$$

Directrix (1) A straight line associated with a conic, from which the shortest distance to any point on the conic maintains a constant ratio with the distance from that point to the focus.

(2) A plane curve defining the base of a cone or cylinder.

Dirichlet's Test for Convergence. If Σu_n converges or oscillates between finite limits and $a_1, a_2, a_3, \ldots$ is a decreasing sequence of positive terms which tends to zero as a limit, then,

$$\Sigma a_n u_n \text{ is convergent.}$$

Discrete. Refers to the property of set of numbers or events being such that there are no intermediate elements in between any two elements. For example, set of integers is discrete whereas set of rational numbers is not discrete.

Discriminant. Refers to the expression $b^2 - 4ac$ in a quadratic equation $ax^2 + bx + c = 0$ which determines the nature of roots. The roots are real, complex, or zero according as the discriminant is positive, negative, or zero.

Disjunction. Symbol: V. In logic, the relationship between two propositions or statements. Disjunction can be either inclusive or exclusive. *Inclusive disjunction* (sometimes called *alternation*) is the one most commonly used in mathematical logic, and can be interpreted as 'one or the other or both'. For two propositions P and Q. P V Q is false if P and Q are both false, and true in all other cases. The more rarely used *exclusive disjunction* can be interpreted as 'either one or the other but not both'. With this definition P V Q is false when P and Q are both true, as well as when they are both false. The truth table definitions for both types of disjunction are shown in the illustration.

P Q	P V Q
T T	F
T F	T
F T	T
F F	F

Disjunction (exclusive)

Disk. Refers to a device that is widely used in computer system to store information. It is a flat circular metal plate coated generally on both sides with a magnetizable substance. Information is stored in the form of small magnetized spots, these are closely packed in concentric *tracks* on the coated surfaces of the disk. The spots are magnetized in one of two directions such that the information is in binary form. The magnetization pattern of a group of spots represents a letter, digit (0 - 9), or some other characters. One disk can store several million characters. Information can be altered or deleted according to requirement by magnetic means. Disks are generally stacked on a common spindle in a single unit called a *disk pack*. Disk packs storing 200 million characters are common.

Information can be recorded on a disk using a special type writer; this method is called *key-to-disk*. The information is fed into a computer using a complex divice known as a *disk unit*. The disk pack is rotated at very high speed in the disk unit. Small electromagnets, called *read-write heads*, move radially in and out over the surfaces of the rotating disks. They extract (read) or record (write) items of information at specified locations on a track, following instructions from the central processor. The time to reach a specified location is very short. This factor, together with the immense storage capacity, makes the disk unit a major backing store in a computer system.

Displacement. Refers to vector quantity used for distance between two positions of a moving object. Because displacement is a vector it has magnitude as well as direction. Displacement may be rectilinear angular, or both.

Distance. Symbol: d. The length of the path between two points. The SI unit is the metre (m). Distance may or may not be measured in a straight line. It is a scalar; the vector form is displacement.

Distance Formula. Refers to the formula for the distance between two points (x_1, y_1) and (x_2, y_2) in Cartesian coordinates. It is :

$$\sqrt{[(x_1 - x_2)^2 + (y_1 - y_2)^2]}.$$

Distance Ratio (Velocity Ratio). For a machine it is the ratio of the distance moved by the effort in a given time to the distance moved by the load in the same time.

Distribution. A term in statistics describing the set of values of a variable and the frequencies of occurrence of each value.

Distribution Function. A mathematical function that gives the probability of finding a system in a particular state or within a range of states. For example, the probability of one of the molecules of a gas having a velocity between v and u + dv in a given direction is Fdv, where F is the distribution function.

Distributive Law. A law concerning the combination of mathematical operations. In arithmetic, for example, multiplication is said to obey the distributive law, or to be distributive, with respect to addition because $a(b + c) = ab + ac$.

Distributive Property. Refers to a law that connects the operations of multiplication and division. According to this law or property of multiplication over addition $a \times (b + c) = a \times b + a \times c$. In Boolean algebra, if it refers to addition over multiplication, $a + (b \times c) = (a + b) \times (a + c)$ whereas in set-theory $A \cap (B \cup C) = (A \cup B) \cup (A \cup C)$.

Divergence. (1) (Div.). Symbol ∇. A vector operator that when operated on a vector point function $\vec{F}(x, y, z)$ yields a scalar quantity as follows. If $\vec{F} = F_1 i + F_2 j + F_3 k$ then

$$\nabla . F = i.\frac{\partial F}{\partial x} + j.\frac{\partial F}{\partial y} + k.\frac{\partial F}{\partial z} = \frac{\partial F_1}{\partial x} + \frac{\partial F_2}{\partial y} + \frac{\partial F_3}{\partial z}.$$

(2) The phenomenon of an infinite series being *divergent*. Opposite to *convergence*.

Dividend. A share in the profits of a company paid to shareholders. The rate of dividend is declared at the company's annual general meeting and will reflect the preceding year's profit. It is usually expressed as a percentage of the par value of the share e.g., a 5% dividend on a 20p share would pay 1p per share). The dividend *yield* of the share is the income it produces expressed as a percentage of its current value (e.g. a 5% dividend on a 20p share that has a current value of 35p is $20/35 \times 5 = 2.86\%$). Dividends are usually paid twice yearly, an interim dividend and a final dividend.

Division. Refers to the binary operation of finding the quotient of two quantities. Division is the inverse operation to multiplication. In arithmetic, the division of two numbers is not commutative

($2 \div 3 \neq 3 \div 2$), nor associative [$(2 \div 3) \div 4 \neq 2 \div (3 \div 4)$] . The identity element for division is one only when it comes on the right hand side ($5 \div 1 = 5$ but $1 \div 5 \neq 5$).

Documentation. Refers to written instructions and comments which give a full description of a computer program. The documentation describes the purposes for which the program can be used, how it operates, the exact form of the inputs and outputs, and how the computer must be operated. It permits the program to be amended when necessary or to be converted for use on different types of machines.

Domain. Refers to a set of numbers or quantities on which a mapping is, or may be, carried out. In algebra, the domain of the function f(x) is the set of values that the independent variable x can take. If, for example f(x) represents taking the square roof of x, then the domain might be defined as all the positive rational numbers.

D Operator. Refers to the differential operator d/dx. The derivative df/dx of a function f(x) is written as Df. This notation is applied in solving differential equations. A second derivative, $d^2 f / dx^2$, is written as D^2f, a third derivative, d^2f/dx^3, as D^3f, and so on. In some ways, the D operator can be considered like an ordinary algebraic quantity, inspite of the fact that it has no numerical value. For example, the differential equation

$$d^2 y/dx^2 + 2x\, dy/dx + dy/dx + 2x = 0$$

or

$$D^2y + 2xD + D + 2x = 0$$

can be factorized to $(D + 2x)(D + 1) = 0$. (D + 2x) then operates on the function (D + 1).

Double Integral. Refers to the result of integrating the same function twice, first with respect to one variable, keeping a second variable constant, and then with respect to the second variable, keeping the first variable constant. For example, if f(x, y) is a function of the variables x and y, then the double integral, first with respect to x and then with respect to y, is :

$$\iint f(x, y)\, dydx$$

This is equivalent to summing f(x, y) over intervals of both x and y, or to calculating the volume bounded by the surface representing f(x, y). The integral is not affected by the order in which the integrations are carried out if they are definite integrals. Another kind of double integral is the result of integrating twice with respect to the same variable. For example, if a car's acceleration a increases with time t in a known way, then the integral

$$\int a\,dt$$

is the velocity (v) expressed as a function of time; the double integral

$$\int\int a\,dt^2 = \int v.\,dt = x$$

where x is the distance travelled as a function of time.

Double Point. Refers to a singular point on a curve where the curve crosses itself or is tangential to itself. There are several types of double point. At a *node* the curve crosses over itself forming a loop. In this case it has two distinct tangents. At a *cusp* it doubles back on itself and has only one tangent. At a *tacnode* two arcs of curve touch each other and have the same tangent, but unlike a cusp, the arcs continue through the singular point to form four arms. An *isolated double point* may also occur. This satisfies the equation of the curve but does not lie on the main arc of the curve.

Drum. Refers to a metal cylinder coated with a magnetizable substance and used in a computer system for storing information. The information is stored in the form of small magnetized spots, which are closely packed in concentric *tracks* around the circumference of the drum. When in use the drum is rotated at high speed. Small electromagnets, known as read-write heads, are fixed in position over each track. They extract (read) or record (write) items of information in particular locations on the track, as directed by the central processor. The time taken to obtain an item of information is extremely short. Drums are now used only for a few special applications in computing.

Doudecimal. Denoting or based on twelve. In a doudecimal number system there are twelve different digits instead of ten. If, for example, ten and eleven were given the symbols A and B respectively, 12 would be written as 10 and 22 as 1A. Doudecimal numbers are of little use, but some duodecimal units (1 foot = 12 inches) are still in use.

Duffing Equation. A differential equation describing the oscillations of a non-linear spring :

$$\frac{d^2y}{dt^2} + a\frac{dy}{dt} + y + by^3 = 0.$$

When a > 0 and b > 0 ('hard spring') the solution is stable even for an arbitrarily large initial disturbance. For a > 0 and b < 0 soft spring') the solution is only stable up to a critical value of the initial disturbance.

Duhamels Theorem. Let L be a linear operator whose coefficients and derivatives do not involve the time variable t, and $\Phi(x, t)$ be the solution of the differential equation

$$L\Phi + A(x)\frac{\partial^2\Phi}{\partial t^2} + B(x)\frac{\partial\Phi}{\partial x} = 0 \quad (0 < x < L, t > 0)$$

having boundary conditions

$$\Phi(x,0) = 0, \frac{\partial\Phi}{\partial t} = 0 \quad (0 < x, L, t = 0)$$

and boundary conditions of the form

$$\alpha\frac{\partial\Phi}{\partial x} + \beta\Phi = b(t) \quad (x = 0, t > 0).$$

such that there are as many homogeneous boundary conditions for x = 0 and/or L as are required for solution.

Then if $\chi(x, t)$ is the solution for the special case of $b(t) = 1$ $t > 0$ the function $\phi(x, t)$ is given by

$$\Phi(x, t) = \chi(x, t)\,b(0) + \int_0^t \chi(x, t - \tau)\,b'(\tau)\,d\tau$$

$$= \frac{\partial}{\partial t}\int_0^t \chi(x, t - \tau)\,b(\tau)\,d\tau \quad (0 < x < L, t > 0)$$

The theorem is used for obtaining the transient response of a dynamical system of any arbitrary force, and has particular application to electrical networks.

of the hyperbola may be written as $x = a \sec\theta$, $y = b \tan\theta$, θ being the *parameter*.

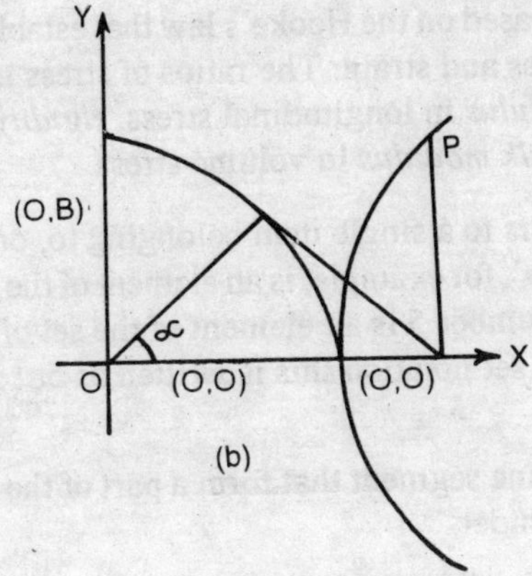

Fig. E-2.

Eccentricity. A measure of the shape of a conic. The ecentricity is the ratio of the distance of a point on the curve from a fixed point (the focus) to the distance from a fixed line (the directrix). For a parabola, the eccentricity is one. For a hyperbola, it is greater than one, and for an ellipse, it is between 0 and 1. A circle has an eccentricity of 0.

Efficiency. Symbol: η. Refers to the ratio of the useful energy produced by a system or device to the energy input. For example, the efficiency of an electric motor is the ratio of its mechanical power output to the electrical power input. There is no unit of efficiency; however efficiency is often quoted as a percentage. In practical systems some dissipation of energy always take place by (friction, air resistance, etc.) and the efficiency is always less than 1. For a machine, the efficiency is the force ratio divided by the distance ratio.

Elastic Collision. Refers to a collision for which the restitution coefficient is equal to one. Kinetic energy is conserved during an elastic collision. In actual practice, collisions are not perfectly elastic because some energy is transferred to internal energy of the bodies.

Elasticity. Refers to the property of a body that resists deformation and brings the body to its original shape and size when the external forces responsible for deformation are removed. The theory of elasticity is based on the Hooke's law that establishes relationship between stress and strain. The ratios of stress to strain are called *Young's modulus* in longitudinal stress, *rigidity modulus* in stear stress and *bulk modulus* in volume stress.

Element. (1) Refers to a single item belonging to, or is a member of a set, 'February', for example, is an element of the set (months in the year). The number 5 is an element of the set of integers between 2 and 10. In set notation this is written as $5\varepsilon\{2, 3, 4, 5, 6, 7, 8, 9, 10\}$.

(2) A line segment that form a part of the curved surface of a cone or cylinder.

(3) A small part of a line, surface, or volume summed by integration.

Eliminant (characteristic, resultant). Refers to the relationship between coefficients that results from eliminating the variables from a set of simultaneous equations. For example, in the equations

$$a_1x + b_1y + c_1 = 0$$
$$a_1x + b_1y + c_1z = 0$$
$$a_2x + b_2y + c_2z = 0$$
$$a_3x + b_3y + c_3z = 0$$

The eliminant is given by the matrix determinant equation

$$\begin{vmatrix} a_1 & b_1 & c_1 \\ a_2 & b_2 & c_2 \\ a_3 & b_3 & c_3 \end{vmatrix} = 0$$

$$a_1x + b_1y + c_1 = 0$$
$$a_2x + b_2y + c_2 = 0$$
$$a_3x + b_3y + c_3 = 0$$

$$\begin{vmatrix} a_1 & b_1 & c_1 \\ a_2 & b_2 & c_2 \\ a_3 & b_3 & c_3 \end{vmatrix} = 0$$

A set of three simultaneous equations and the eliminant given by the corresponding matrix determinant equation.

Elimination. Means removing one of the unknowns in an algebraic equation, for example, by the substitution of variables or by cancellation.

Ellipse. A closed curve having the shape of an elongated *circle. The sum of the distances from any point on the circumference to each of two fixed points, known as the foci, is a constant. In the Cartesian *coordinate system its equation is $(x - h)^2 / a^2 + (y - k)^2/b^2 = 1$, where (h, k) is the centre of the ellipse and a and b are the largest and shortest radii, which are parallel to the coordinate axes. In the case of a and b being equal, the ellipse becomes a *circle of radius a. The ellipse is one of a family of curves known as *conic sections.

Ellipsoid. Refer to a solid whose plane cross-sections parallel to the three coordinate axes are ellipses. It has three axes of symmetry. The equation $\frac{x^2}{a^2} + \frac{y^2}{b^2} + \frac{z^2}{c^2} = 1$ represents the standard cartesian equation of ellipsoid having 2a, 2b, 2c as lengths of its three axes. If b = c the ellipsoid is of *prolate* shape and is formed by the revolution of the ellipse $\frac{x^2}{a^2} + \frac{y^2}{b^2} = 1$ about its major axis. If a = b, the ellipsoid is of *oblate* shape and is formed by the revolution of the ellipse $x^2/a^2 + z^2/c^2 = 1$ about its minor axis. If a = b = c the ellipsoid is the sphere $x^2 + y^2 + z^2 = a^2$.

Encke Roots. Let the roots of

$$x^2 + a_1 x + a_2 = 0$$

be x_1 and x_2 where $|x_2| > |x_1|$. Thewe roots with their signs changed, are known as the Encke roots of the equation.

Energy. Refers to capacity of a system to do work. Energies can be broadly categorised as *kinetic energy* and *potential energy*.

Engineering. The systematic application of scientific knowledge to the design, creation, and use of structures and machines. Engineering has its roots in the consructions in classical times by military engineers of harbours, roads, aqueducts, tunnels, canals, and siege machines. Civil engineering arose from the study and design of such static structures as bridges, dams, buildings, etc., whereas mechanical engineering is concerned with dynamical systems, such as machinery and engines. Other important branches of engineering are electrical engineering, aeronautical engineering, and chemical engineering.

Enlargement. Refers to a geometrical projection which produces an image larger (smaller if the scale factor is less than 1) then, but similar to, the original shape.

Entailment. In logic refers to the relationship which holds between two (or more) propositions when one can be deduced from the other. If conclusion C is deducible from premisses A and B, A and B are said to *entail* × C.

Epicycle. Refers to a circle that rolls around the circumference of another, tracing out an epibycloid.

Epicycloid. Refers to the plane curve traced out by a point on a *epicycle* rolling along the outside of another fixed circle. For example, if a small cog wheel turns on a larger stationary wheel, then a point on the rim of the smaller wheel traces out an epicycloid. In a two-dimensional Cartesian-coordinated system which has a fixed circle of radius a centered at the origin and another of radius b rolling around the circumference, the epicycloid is a series of continuous arcs which move away from the first circle to a distance 2b and then return to touch it again at a cusp point at which the next arc begins. The epicycloid has only one arc if a = b, two if a = b/2, and so on. If the angle between the radius from the origin to the moving point of contact between the two circles is θ, the epicycloid is defined by the parametric equations :

$$x = (a + b) \cos \theta - a \cos[(a + b)\theta/a]$$

$$y = (a + b) \sin \theta - a \sin [(a + b)\theta/a]$$

Equality. Refers to the property of two quantities being equal. This is generally expressed in the form of an equation. Two complex numbers will be equal if their real parts and imaginary parts are respectively equal. Thus, $a + ib = c + id$ if $a = c$ and $b = d$.

Equinox. Either of the two points at which the *ecliptic intersects the celestial equator. The ecliptic represents the apparent annual path of the sun around the celestial sphere. The sun crosses the celestial equator from S to N at the **vernal** (or spring) **equinox** (21 March). It crosses from N to S at the **autumnal equinox** (23 Sept.). At the equinoxes the centre of the sun is above and below the horizon for an equal time, and night and day are then of almost equal duration.

Equation. Refers to a statement of equality. If the equality is true only for some values of the unknown quantities involved in the equation, then it is called conditional equation and the = sign is used; if the equality is true for all values of the unknown quantities involved, the equation is known as an *identity* and the ≡ sign is used.

Equation of Motion. Refers to the equations that describe the motion of an object. If we consider the linear motion only and acceleration to be constant, then the equations can be written as

$$v = u + at$$

$$s = ut + \tfrac{1}{2} at^2$$

or

$$s = (u + v)\, t/2$$

$$v^2 = u^2 + 2as$$

where u and v denote the initial and final velocities respectively in a time interval t, s is the distance travelled in time t and a is the constant aceleration. a is repladced by — a if it is retardation.

Equator. Refers to the great circle on the surface of the earth which is equidistant from the north pole and south poles. The equator divides the earth's surface into the northern and southern hemispheres. The plane wherein equator lies is known as *equitorial plane*.

Equilibrium. Refers to the state of a body or particle when the resultant of all the forces acting on it is zero and there is no torque acting on it. The equilibrium is *stable* or *unstable* according as when slight displacement is given to the body, the body tends to return to or move away from its original position respectively, under the action of force acting on the body.

Equipotential, Equipotential Surface. Refers to a surface on which a potential function U maintains a constant value. More generally, if U is any point function, then an equipotential surface relative to U is a surface on which U is constant.

Equivalence, Equivalence Class. If an *equivalence relation* is defined on a set, then the set can be separated into classes by the convention that two elements belong to the same class if and only if they are equivalent. These classes are *equivalence classes.* Two equivalence classes are identical if they have a common element. Each element belongs to one of the equivalence classes. E.g., if one says that a is equivalent to b if a - b is a rational number, this is an equivalence relation for the real numbers: the equivalence class that contains a number a is then the class that contains all numbers that can be obtained by adding rational numbers to a. If a set T has been divided into nonoverlapping subsets, then an equivalence relation can be defined by letting "x is related to y" mean that x and y are in the same subset.

Equivalence of Propositions. An *equivalence* is a proposition obtained from two given propositions by connecting them by "*if and only if*". An equivalence is true if both propositions are true, or if both are false. The proposition "For all triangles x, x is equilateral if, and only if, x is equiangular" is true, since any particular triangle is either both equilateral and equiangular, or is neither equilateral nor equiangular. The equivalence formed from propositions p and q is generally denoted by $p \leftrightarrow q$ or $q \equiv q$. The equivalence $p \leftrightarrow q$ is same as the statements "p is a *necessary* and *sufficient* condition for q." or "p if, and only if, q" it is equivalent to the *conjunction* of the imnplications $p \rightarrow q$ and $q \rightarrow p$. Propositions p and q are *equivalent* if $p \leftrightarrow q$ is true. An equivalence is also known as a *biconditional* statement (or proposition). Two statements are *logically equivalent* when they are equivalent due to their logical form rather than due to mathematical content. e.g., $\sim (p \wedge q)$ is equivalent to $(\sim q) V (\sim q)$ whatever the statements p and q.

Equivalence Relation. Refers to a relation between elements of a set which satisfies three conditions : (a) *Reflexive, aRa,* (b) *Symmetric,* that is, if *aRb* then *bRa,* (c) *Transitive,* that is, if *aRb* and *bRc* then *aRc,* where a, b and c all belong to the set. For example, congruence in geometry, equality in algebra.

Eratosthenes, Sieve of. The prime numbers not greater than N can be found by writing all numbers up to N and then removing those numbers that are multiples of 2, 3, ...etc., continuing until all the primes not greater than $\sqrt{N}$ have been removed.

Error. For any set for measures of the same quantity there is a mean that is considered to be a theoretical 'true value' i.e., a unique value that exists only as a concept. Error is the difference between a particular measure and a set of measures. In statistics it can be the variation because of factors beyond control or because of sampling error i.e., owing to the arbitrary nature of the sample chosen to represent the population.

Estimate. Refers to (1) A rough calculation, generally involving one or more approximations made to give a preliminary answer to a problem.

(2) A statement indicating the approximate cost for carrying out a specified job e.g., painting a house or repairing a car. The estimate should state the assumptions that have been made and the extent to which the total sum is likely to differ if the assumptions are invalid.

Euclidean Algorithm. Refers to a method of finding the highest common factor of two positive integers. The smaller number is divided into the larger. The remainder is then divided into the smaller number, obtaining a second remainder. This second remainder is then divided into the first remainder, to give a third remainder. This is divided into the second, and so on, until a zero remainder is obtained. The remainder preceding this is the highest common factor of the two numbers. For example, the numbers 54 and 930. Dividing 54 and 930 gives 17 with a remainder of 12. Dividing 12 into 54 gives 4 with a remainder of 6. Dividing 6 into 12 gives 2 with a remainder of 0. Thus 6 is the highest common factor of 54 and 930.

Euclidean Geometry. The study of ordinary two-or-three-dimensional space with certain Euclidean axioms : e.g., the shortest distance between two points is a straight line; two halves of equal objects are equal, etc. It can also be used to mean the study of Euclidean space of any number of dimensions. This space may be defined as all sets of n numbers $(x_1, x_2, ..., x_n)$, where the distance $\rho(x, y)$ between $x = x(x_1, x_2, ...,x_n)$ and $y = y(y_1, y_2, ..., y_n)$ is defined as

$$\rho(x, y) = \left[\sum_{i=1}^{n} | x_i - y_i |^2 \right]^{1/2}$$

The distance is real or complex according as x is a real or complex number.

Euclid's Alogrithm (for the Highest Common Factor). If X and Y are polynomials of degree n and m respectively, where $n \geq m \geq 1$, then their highest common factor can be found as follows: divide X by Y and obtain a remainder R_1 where R_1 is of degree $m_1 < m$. If $R_1 \neq 0$, divide Y by R_1 to obtain a remainder R_2 of degree $m_2 < m_1$ and divide R_1 by R_2. Continue unit some $R_i = 0$, when R_{i-1} is the highest common factor, or until R_i is a constant, in which case the highest common factor is 1.

Euler Characteristic. Refers to a topological property of a curve or surface. For a curve, the Euler characteristic is the number of vertices minus the number of closed continuous line segments between. For example, any polygon has an Euler characterises of zero. For a surface, the Euler characteristic is equal to the number of vertices plus the number of faces minus the number of edges. For example, a cube has an Euler characteristic of 2, and a cylinder, a Mobius strip, and a Klein bottle have an Euler characteristic of zero.

Eulerian Angles. In considering the kinematics of solid bodies, a set of angles serving both to define the orientation of the solid body and as independent variables in the equations of motion.

If x, y and z are the three axes fixed in the body, and X, Y, Z represent a system of axes whose direction in space is fixed, θ is the angle between the Z and z axes. The line in which the xy plane cuts the XY plane is called the line of nodes and the angle between this and the x axis is termed Ψ. The angle from the X axis measured in the positive direction Z to the line of nodes is called ϕ. (Alternative definitions of the Eulerian angles are sometimes used).

Euler's Formula. (1) Refers to (for polyhedra). The formula that combines the number of vertices v, faces f, and a edges e in a polyhedron, that is :

$$v + f - e = 2$$

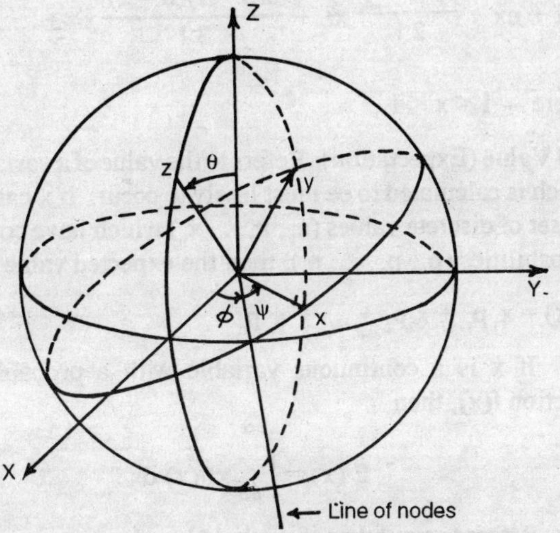

Fig. E-3.

For example, a cube has eight vertices six faces and twelve edges:

$$8 + 6 - 12 = 2$$

Making use of the theorem it can be shown that there are only five regular polyhedrons.

(2) The difinition of the function $e^{i\theta}$ for any real value of θ, where i is the square root of —1, as

$$e^{i\theta} = \cos\theta + i\sin\theta.$$

Any complex number $z = x + iy$ can be written in this form. $x = r\cos\theta$ and $y = r\sin\theta$ are real, with r and θ representing z on an Argand diagram. Note that putting $\theta = \pi$ given $e^{i\pi} = -1$ and $\theta = 2\pi$ gives $e^{2\pi t} = 1$.

Expansion. Refers to the process of expressing a quantity as a sum of a series of terms. Many times a function can be written as a convergent infinite series. The function, in such a case, can be approximated to any required degre of accuracy by taking the sum of a certain number of terms from the series. The expansion of $(1 + x)^n$ is

$$1 + nx + \frac{n(n-1)}{2!}x^2 + \frac{n(n-1)(n-2)}{3!}x^3 + \ldots.$$

where $-1 < x < 1$.

Expected Value (Expectation). Refers to the value of a variable quantity which is calculated to be most likely to occur. If x can take any of the set of discrete values $(x_1, x_2, \ldots x_n)$ which have corresponding probabilities $(p_1, p_2, \ldots p_n)$, then the expected value

$$E(x) = x_1 p_1 + x_2 p_2 + \ldots + x_n p_n.$$

If x is a continuous variable with a probability density function f(x), then

$$E(x) = \int_{-\infty}^{\infty} xf(x)\, dx$$

Exponent. Refers to a number of symbol placed as a superscript after an expression to show the power to which it is raised. For example, x is an exponent in y^x and in $(ay = b)^x$.

The laws of exponent are used for combining exponents of numbers as follows :

Multiplication :

$$x^a x^b = x^{a+b}$$

Division :

$$x^a / x^b = x^{a-b}$$

Power of a power :

$$(x^a)^b = x^{ab}$$

Negative exponent :

$$x^{-a} = 1/x^a$$

Fractional exponent :

$$x^{a/b} = b\sqrt{x^a}$$

A number raised to the power zero is equal to 1 ; i.e., $x^o = 1$.

Exponential. Refers to a function of quantity which changes as the power of another quantity. In $y = 4^x$, y is said to vary exponentially with respect to x. The function e^x (or expx), where e is the base of natural logarithms, is the exponential of x.

The infinite series

$$1 + x + x^2/2\,! + x^3/3\,! + \ldots + x^n/n\,! + \ldots$$

is equal to e^x and is called the *exponential series.* The exponential form of a complex number is

$$re^{i\theta} = r\,(\cos\theta + i\sin\theta).$$

Exponential Function. A mathematical function of the general form Ae^{Bx} where A and B are constants and e is the base of natural logarithms. Bx is called the exponent; if the exponent is a complicated expression, f(x) say, then the function is often written as exp [f(x)]. The function e^x has the important property that its differential is always equal to the function itself.

Exponential Series. Refers to the infinite power series that is the expansion of the function e^x,

$$1 + x + x^2/2\,! + x^3/3\,! + \ldots\ldots.$$

$$x^n/n\,! + \ldots\ldots$$

This series is convergent for all real number values of the variable x.

Replacing x by —x gives an alternating series for e^{-x}:

$$1 - x + x^2/2\,! - x^3/3\,! \ldots\ldots.$$

Series for sinh x and cosh x can be obtained by combining series for e^x and e^{-x}.

Exterior Angle. Refers to the angle formed on the outside of a plane figure between the extension of one straight edge beyond a vertex, and the outer side of the other straight edge at that vertex. In a triangle, the exterior angle at one vertex equals the sum of the angles on the inside of the other two vertices; i.e., the sum of the interior opposite angles.

Extrapolation. Refers to the process of estimating the value of a function or quantity outside a known range of values. For example, if the speed of an engine is controlled by a lever, and depressing the lever by two, four, and six centimetres gives speeds of 20, 30 and 40 revolutions per second respectively, then one can extrapolate from this information and make the assumption that depressing it by a further two centimetres will increase the speed to 50 revolutions per second. Extrapolation can also be done graphically; for example, a graph can be drawn over a known range of values and the resulting curve extended. The further from the known range this line is taken, the greater will be the uncertainty in the extrapolation. The case wherein the graph of the behaviour is straight line (as in the example above) is a *linear extrapolation.*

F

Factorization. Refers to the process of changing any algebraic or numerical expression from a sum of terms into a product of factors. A number of expression is called a *prime* is no factors can be obtained of that.

Factor Theorem. (x — a) is a factor of the polynomial f(x) in x if and only if f(z) = 0. For example, if

$$f(x) = x^2 + 5x + 6 = 0,$$

$$f(-2) = f(-3) = 0$$

so that(x + 2) and (x + 3) are factors of f(x).

Fahrenheit Scale. Refers to a scale of temperature on which the freezing point of water and its boiling point are doivided into one hundred and eighty parts. On this scale water freezes at 32°F and boils at 212° F. The formula for converting centigrade temperature to Farenheit scale is

$$\frac{C}{5} = \frac{F - 32}{9}$$

where C is the temperature on Centigrade scale and F is temperature on Fahrenheit scale.

False, Method of False Position. Refers to a method for approximating the roots of an algebraic equation. Consists of making a fairly close estimate, say r, then substituting (r + h) in the equation, dropping the terms in h of higher degree than the first (since they are relatively small), and solving the resulting linear equation for h. This process is repeated, using the new approximation (r + h) in place of r.e.g., the equation $x^3 - 2x^2 - x + 1 = 0$ has a root near 2 (between 2 and 3). Hence, we subsitute (2+h) for x. This gives (when the terms in h^3 and h^2 have been dropped) the equation

3h — 1= 0; whence h = 1/3. The next estimate will then be 2 + 1/3 or 7/3 and so on.

Family. Refers to a set of related curves or figures. For example, a family of parallel straight lines is represented by the equation $y = 2x + k$.

Family of Curves. A set of curves having same geometrical pattern. A family of curves is represented by an equation having one or more parameters. Different values of the parameters give different members of the family. For example, the equation $x^2 + y^2 = c^2$, represents a family of concentric circles. Different values of c give different curves of the family.

F Distribution. Refers to the satistical distribution followed by the ratio of variances s_1^2 / s_2^2 of pairs of random samples, size n_1 and n_2 taken from a normal distribution. It is generally used to compare different estimates of the same variance.

Fibonacci Ratios. Refers to the ratios of any two consecutive numbers from the *Fibonacci sequence,* i.e., the ratios 1/1, 1/2, 2/3, 3/5, 5/8, 8/13,......The limit of the sequence of these ratio is known as the *golden ratio* whose value is $(\sqrt{5} - 1)/2$. The distribution pattern around the stem of a plant of successive leaves very much fits Fibonacci ratios.

Figure. Refers to a shape formed by a combination of points, lines, curves, or surface. Circles, squares, and triangles are plane figures, whereas Spheres, cubes and pyramids are solid figures.

Finite Set. Refers to a set which has a fixed countable number of elements. For example, the set of 'months in the year' has 12 members and is thus a finite set.

First Order Differential Equation. A differential equation wherein the highest derivative of the dependent variable is a first derivative.

Flowchart. Refers to a diagram where on can be represented the major steps in a process used, say, in industry, or a problem to be investigated, or a task to be carried out. A flowchart is built up from a number of boxes joined by arrowed lines. The boxes, of different shapes, have a label attached showing for example the operation or calculation to be done at each step. At a decision box a question is asked. The answer, either yes or no, gives which of two possible paths to take. Computer programs are generally written by first drawing a flowchart of the problem or task in hand.

Fly Wheel. Refers to a big wheel that is used in mechanical devices, having large moment of inertia. Energy is used to rotate the wheel at a high speed. Because of large moment of inertia the wheel continues to rotate at a constant speed. The wheel acts like an energy storage device.

Force. Symbol : f. It is that agency which tends to change an object's momentum. Force is a vector; the unit is the newton (N). In SI this unit is so defined that :

$$F = d(mv)/dt$$

from Newton's second law.

Force Ratio. (mechanical advantage) For a machine refers to the ratio of the output force (load) to the input force (effort). It has no unit; the ratio is, however, sometimes given as a percentage. It is quite possible for force ratios far greater than one to be obtained. Indeed many machines are designed for this so that a small effort may overcome a large load. However, the efficiency cannot be greater than one and a large force ratio implies a large distance ratio.

Formula. Refers to a general expression that can be used for several different values of the quantities in question. For example, the formula for the area of a circle is πr^2, where r is the radius.

Fortran (formula translation). A computer-programming language. It is a high-level language used for representing mathematical formulae in a form that can be processed by a computer. Widely used among scientists and engineers, it is also used in the business community.

Foucault Pendulum. Refers to a simple pendulum of a heavy bob on a long string. The period is large and the plane of vibration rotates slowly over a period of time due to the rotation of the Earth below it. The apparent force causing this movement is the Coriolis force.

Four-Colour Problem. A problem in topology relating to the division of the surface of a sphere into regions. The name comes from the colouring of maps. It seems that in colouring a map it is not essential to use more than four colours to distinguish regions form each other. Two regions having a common line boundary between them need different colours, but two regions meeting at a point do not. On the surface of a torus only seven colours are necessary to distinguish regions.

Free Oscillation (Free vibration). An oscillation at the natural frequency of the system or object. Thus a pendulum can be forced to swing at any frequency; it will swing freely at only one, that depends on its length and mass.

French Curve. A drawing instrument made up of a rigid piece of plastic of metal sheeting with curved edges. The curvature of the edges differs from being almost straight to very right curves, so that a part of the edge can be chosen to guide a pen or pencil along any desired curvature. Another instrument that can be used for the same purpose is made up of a deformable strip of lead bar cut into short sections and surrounded by a thick layer of plastic. This is bent to form any required curvature.

Frequency. Symbol : f, v. Refers to the number of cycles per unit time of an oscillation, (e.g., a pendulum, vibrating system, wave, alternating current, etc.). The unit is the hertz (Hz). The symbol f is used for frequency, although v is often used for the frequency of light or other electromagnetic radiation.

Angular frequency (ω) is related to frequency by $\omega = 2\pi f$.

Frequency Curve. Refers to a smoothed frequency polygon for data that can take a continuous set of values. As the amount of data is increased and the size of class interval decreased, the frequency polygon more closely approximates a smooth curve. Relative frequency curves are smoothed relative frequency polygons.

Frequency Function. For a variable x that has only a finite or denumerably infinite number of values, the **absolute frequency function** is the function f for which $f(x_i)$ is the absolute frequency of x_i; the **relative frequency function** is the function g for which $g(x_i)$ is the relative frequency of x_i. For a *random variable* having possible values $X_1, X_2, ..., X_n$, the **frequency function** is the function p for which $p(X_i)$ is the probability of X_i; often this is known as a probability function, but strictly it is the restriction of the probability function to the events X_1, X_2, ..., X_n.

Friction. Refers to a force opposing the relative motion of two surfaces in contact. In fact, each surface applies a force on the other in the opposite direction to the relative motion; the forces are parallel to the line of contact. The exact reasons of friction are still not fully understood. If probably results from minute surface roughness, even on apparently 'smooth' surface. Frictional forces do not depend on the area of contact. Presumably lubricants act by separating the surface. For friction between two solid surfaces, *sliding friction* (or *kinetic friction*) opposes friction between two moving surfaces. It is less than the force of *static* (or *limiting*) *friction,* that opposes slip between surfaces that are at rest. *Rolling friction* take place when a body is rolling on a surface : here the surface in contact is constantly changing. Frictional force (F) is proportional to the force holding the bodies together (the 'normal reaction' R). The constants of proportionality (for different cases) are known as *coefficients of friction* (Symbol: μ) :

$$\mu = F/R.$$

The *laws of friction* are stated as :

(1) The frictional force is independent of the area of contact for the same force holding the surfaces together.

(2) The frictional force is proportional to the force holding the surfaces together. In sliding friction it is independent of the surfaces together. In sliding friction it is independent of the relative velocities of the surfaces.

Function (mapping). Refers to any defined procedure that relates one number, quantity, etc., to one or more others. In algebra, a function of a variable x is generally written as f(x). If two variable quantities, x and y are related by the equation $y = x^2 + 2$, for example, then y is a function of x or $y = f(x) = (x)^2 + 2$. The function here means 'square the number and add two'. x is the *independent variable* and y is known as the *dependent variable*. The inverse function — is one that expresses x in terms of y in this case—would be $x = +\sqrt{(y - 2)}$, that might be written as x = g(y).

A function may be regarded as a relationship between the

elements of one set (the *range*) and those of another set (the *domain*). For each element of the first set there is a corresponding element of the second set into which it is 'mapped ' by the function.

Functional Analysis. Refers to an extension of the study of functions of real or complex variables, wherein every set of function is associated with a unique number. This correspondence is known as *functional*. Thus $\int_b^a f(x)\,dx$, can be interpreted as a functional on the set of all function integrable on (a, b). The value of any functional can be obtained only by considering all possible behaviours of the function.

Fundamental Operators. Refers to operations such as addition, subtraction, multiplication, division; involution, evolution; rules of indices, logarithms; differentiation, integration; inversion, transformation, translation, rotation, reflection. Out of these the first four operations are fundamental operations of arithmetic.

Fundamental Theorem of Algebra. Every polynomial equation of the form

$$a_0 z^n + a_1 z^{n-1} + a_2 z^{n-2} + \dots + a_{n-1} z + a_n = 0$$

where a_0, a_1, a_2, ... a_n are all complex numbers, has atleast one complex root.

G

Gallon. Refers to the British standard unit of capacity for measuring volume of liquids and is defined as the space occupied by ten pounds of water at N.T.P.(normal temperature and pressure). It is equal to 4.547 litres.

Game Theory. The branch of mathematics that seeks to analyse and solve problems arising in economic, business, or military situations on the assumption that each participant adopts strategies that will maximize gain (payoff) and minimize loss, as in playing a game. Worked out originally by John von Neumann and Oskar Morganstern (1907—), it was successfully used in World War II in an analysis of submarine warfare.

Distinction is made between games for one player (patience, solitaire) in which no conflict arises, games for two players or teams (chess, football) in which conflict is an essential aspect of strategy, and games in which there are more than two participants (poker, roulette) and one person's gain is not directly reflected by another person's loss. When chance is important, information is incomplete, or participants' goals are obscure, the theory is more complex.

Gamma Function. Gamma of a number n is defined as

$$n = \int_0^\infty x^{n-1} e^{-x}\, dx$$

When n is a positive integer $|\bar{n} = (n-1)!$ It is easy to see the $|1 = |2 = 1$, $|n+1 = n|n$ and $|\tfrac{1}{2} = \sqrt{\pi}$.

General Form (or an equation). Refers to a formula which defines a relationship between variables but does not specify values for constants. For example, the general form of a polynomial equation in x is

$$ax^n + bx^{n-1} + cx^{n-2} + \ldots = 0$$

a, b, c, etc., are constants and n is the highest integer power of x, known as the *degree* of the polynomial, similarly, the general form of a quadratic equation is

$$ax^2 + bx + c = 0$$

Geodesic. Refers to a line on a surface between two points which acts as the shortest distance between these two points.

Geometric Distribution. Refers to the distribution of a number of independent Bernoulli trials, before getting a success. For example, the distribution of number of times a coin has to be tossed before getting a head. If p and q are the probabilities of success and failure respectively then the probabilities of getting of success in k trials is $p\ (x = k) = q^{k-1}\,p$.

Geometric Means. The terms in a geometric progression lying between the first and the last. In between two numbers a and b we can insert n such means. In this case b becomes the (n + 2)th term and if r is the common ratio then $b = ar^{n+1}$ thus, the n geometric means in between a and b are given by

$$a(b/a)^{\frac{1}{n+1}},\quad a(b/a)^{\frac{2}{n+1}}, \ldots\ldots\quad a(b/a)^{\frac{n}{n+1}},$$

If n = 1 then the geometric mean of a and b is $\sqrt{ab}$. The geometric mean of n quantities $a_1, a_2, .., a_n$ is

$$(a_1 . a_2 \ldots a_n)^{1/n}$$

Geometric Sequence (geometric progression). Refers to a sequence wherein the ratio of each term and the one after it is constant, for example, {1, 3, 9, 27,.....}. The general formula for the nth term of geometric sequence is $u_n = ar^n$. The ratio is known as the *common ratio*. In the example the first term, a is 1, the common ratio. r, is 3, and u_n equals 3^n. If a geometric sequence is convergent, lies between 1 and —1(exclusive) and the limit of the sequence is 0. That is, u_n approaches zero as n becomes infinitely large.

Geometric Series. Refers to series wherein the ratio of each term to the one after it is constant, for example, 1 + 2 + 4 + 8 + 16... The general formula for a geometric series is

$$S_n = a + ar + ar^2 + \ldots. + ar^n = a(r^n - 1)/(r - 1)$$

In the example, the first term, a, is 1, the *common ratio,* r, is

2, nth term ar^n is given by 2^n. If r is greater than 1, the series will not be convergent. If $-1 < r < 1$ and the sum of all the terms after the nth term can be made as small as needed by making n large enough, then the series is convergent. This means that there is a finite sum even when a is infinitely large. The *sum to infinity* of a convergent geometric series is $a/(1 - r)$.

Geometry. A branch of mathematics concerned with the properties of space and shapes. In Euclidean geometry the space corresponds to common motions of physical space and the shapes are idealizations of the common shapes that occur in real life. Other branches of geometry include non-Euclidean geometry, such as the geometry of the surface of a sphere; Riemannian geometry, which is used in relativity theory; and analytic geometry, in which algebra is used to solve geometrical problems.

Golden Ratio. Refers to the ratio $\lambda : 1$ where λ is the positive root of the equadratic $x^2 - x - 1 = 0$. Thus the golden ratio is $\frac{\sqrt{5}+1}{2}$.

It can also be written as λ^2/λ or $\frac{\sqrt{5}-1}{3-\sqrt{5}}$. This ratio is the limit of the sequence

$$\frac{1}{1}, \frac{2}{1}, \frac{3}{2}, \frac{5}{3}, \frac{8}{5}, \frac{13}{8}, \ldots\ldots$$

It's approximate value is 1.618. It comes from *golden section.*

Golden Section. Refers to the division of a line of length 1 into two length is a and b such that $1/a = a/b$, that is $a/b = (1 + \sqrt{5})/2$. Proportions based on the golden section are specifically pleasing to the eye and found in many paintings, buildings, designs, etc.

Governor. Refers to a mechanical device used to control the speed of a machine. One type of simple of governor is made up of two loads attached to a shaft such that as the speed of rotation of the shaft increases, the loads move further outwards from the centre of rotation, while still remaining attached to the shaft. As they move outwards they operate a control which reduces the rate of fuel or energy input to the machine. As they reduce speed and move inwards they increase the fuel or energy input. Thus, on the principle of negative feedback, the speed of the machine is kept fairly constant under changing conditions of load.

Grad (gradient) Symbol. ∇. Refers to a vector operator which for any function f(x, y,) has components in the x, y and z directions equal to the partial derivatives with respect to x, y and z in that order. It is defined as :

$$\text{grad } f = \nabla f = i\partial f/\partial x + j\partial f/\partial y + k\partial f/\partial z$$

where i, j and k are the unit vectors in the x, y and z directions. In physics, ∇F is generally used to describe the spatial changes in the magnitude of a force F in, a magnetic or gravitational field. It is vector having the direction wherein the rate of change of F is a maximum, if such a maximum exists. In the Earth's gravitational field this would be radially towards the centre of the Earth (downwards). In a magnetic field, ∇ F would point along the lines of force.

Gradient (slope). In rectangular Cartesian coordinates refers to the rate at which the y -coordinate of a curve or a straight line changes with respect to the x coordinate. The straight line $y = 2x + 4$ has a gradient of + 2; y increases by two for every unit increase in x. The general equation of a straight lines is $y = mx + c$, where m is the gradient and c is a constant (0, c) is the point at which the line cuts the y-axis i.e., the intercept. If m is negative, y decreases as x increases.

For a curve, the gradient changes continuously; the gradient of the straight line that is a tangent to the curve at that point. For the curve $y = f(x)$. The gradient is the derivative dy/dx. For example, the curve $y = x^2$ has a gradient which is given by dy/dx $= 2x$, at any particular value of x.

Graph (1) Refers to a drawing that shows the relationship between numbers or quantities. Graphs are generally drawn with coordinate axes at right angles. For example, the heights of children of different ages can be shown by making the distance along a horizontal line represent the height in metres. A point marked on the graph ten units along and 1.5 units up represents a ten-year-old who is 1.5 metres tall. Similarly, graphs are used to give a geometric representation of equations. The graph of $y = x^2$ is a parabola. The graph of $y = 3x + 10$ is a straight line. Simultaneous equations can be solved by drawing the graphs of the equations, and finding the points at which they cross. For the two equations above the graphs cross at two points: $x = -2$, $y = 4$ and $x = 5$ f, $y = 25$.

There are different types of graph. Some, like the histogram and the pie chart, are used to display numerical informaion in a form which is simple and quickly understood. Some like conversion graphs, are used in analysing the results of a scientific experiment.

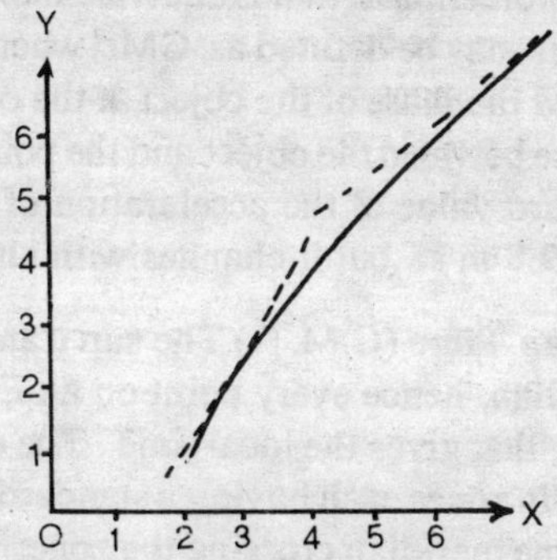

Fig. G-1.

(2) (Topology). A network of lines and vertices.

The gradient of the curve at the point(2, 2) is 2, and at the point (5, 5)is 1/2.

Gravitation. The law of attraction discovered by Issac Newton: According to this law $F = G\,m_1\,m_2/d^2$ where F is the force of attraction between to masses m_1 and m_2 and d is the distance between their centroids. G is the constant of proportionality called *gravitational constant.* It was experimentally established that the force of gravitation between two mass of 1 gram distance 1 cm apart is 6.67×10^8 dynes, and the value of $G = 6.67 \times 10^{-8}$.

Gravitational Constant. Symbol : G. Refers to the constant of proportionality in the equation that expresses Newton's law of universal gravitation:

$$F = Gm_1m_2/r^2$$

where F is the gravitational attraction between two point masses m_1 and m_2 separated by a distance r. The value of G is 6.67×10^{-11} $Nm^2\,kg^{-2}$. It is considered as a universal constant, although it has been suggested that the value of G may be changing slowly due to the expansion of the Universe.

Gravitational Field. Refers to the region of space wherein one body attracts other bodies due to their mass. Two escape from this field a body has to be projected outwards with a certain speed (the *escape speed*). The strength of the gravitational field at a point is equal to the ratio force/mass, that is equivalent to the acceleration of free fall, g. This may be defined as GMr^2 where G is the gravitational constant, M the mass of the object at the centre of the field, and r the distance between the object and the point under consideration. The standard value of the acceleration of free fall at the Earth's surface is 9.8 m s^{-2},but it changes with altitude, (i.e., with r^2).

Greenwich Mean Time (G.M.T.) The sun transists through all points on a meridian, hence every point on an E-W line has a different noon time, that gives the local time. The earth can be considered as divided in zones each having a standard time based on the local time of some meridian crossing the zone. Thus, each hour covers a time extending across 15^0 of longitude. Noon at Greenwich (London) is considered as noon in the British Isles, Portugal, Morocco and a part of West Africa. Noon in Central Europe is at 1.00 p.m. G.M.T. (13.00 hours).

Group. A set having certain additional properties :

(1) In a group there is a binary operation for which the elements of the set can be related in pairs, giving results which are also members of the group (the property of *closure*). For example, the set of all positive and negative numbers and zero form a group under the operation of addition. Adding any member to any other gives an element that is also a member of the group; e.g., 3 + (—2) = 1 etc.

(2) There is an identity element for the operation — i.e., an element which combined with another — leaves it unchanged. In the example the identity element is zero: adding zero to any member leaves it unchanged; 3 + 0 = 3, etc.

(3) For each element of the group there is another element which is its inverse. Combining an element and its inverse leads to the identity element. In the example, the number + 3 has an

inverse —3 (and vice versa) ; thus + 3 + (—3) = 0.

(4) The associative law hold for the members of the group. For the example :

$$2 + (3 + 5) = (2 + 3) + 5.$$

Any set of elements that obey the above rules forms a group. Note that the binary operation need not be addition. *Group theory* is important in many branches of mathematics—for example, in the theory of roots of equations. It is also quite useful in diverse branches of science. In chemistry, group theory is used in describing symmetries of molecules to determine their energy levels and explain their spectra. In physics, certain elementary particles can be classified into mathematical groups according to their quantum numbers (this led to the discovery of the omega-minus particle as a missing member of a group). Group theory has also been applied to linguistics.

H

Harmonic Mean (H). Refers to Harmonic mean of two numbers a and b is given by $H = \frac{2ab}{a+b}$ and that of n numbers $x_1, x_2, \ldots, x_n$ by

$$H = \frac{n}{\sum \frac{1}{x_i}} = \frac{n}{\frac{1}{x_1} + \frac{1}{x_2} + \cdots \frac{1}{x_n}}$$

Harmonic Motion. Refers to a regularly repeated sequence which can be expressed as the sum of a set of sine waves. Each component sine wave represents a possible simple harmonic motion. The complex vibration of sound sources (with fundamental and overtones), for example, is a harmonic motion, as is the sound wave produced.

Harmonic Sequence (Harmonic Progression). Refers to an ordered set of numbers, the reciprocals of which have a constant difference between them; for example, {1, 1/2 1/3, 1/4,....1/n}. The reciprocals of the terms in a harmonic sequence form an arithmetic sequence, and vice versa.

Harmonic Series. Refers to the sum of the terms in a harmonic sequence; for example : 1 + 1/2 + 1/3 +1/4 +...

Hecto. Symbol: h. A prefix denoting 10^2. For example, 1 hectometre (hm) = 10^2 metres (m).

Height. Means. **Altitude.** The vertical distance of a point above the base line or plane. For example, the perpendicular distance from the base of a triangle of the opposite vertex.

Helix. Refers to a space curve in the shape of a spiral. It is known as cylindrical, cerical, or spherical if it lies on a cylinder, cone, or a sphere respectively. For example, the shape of the thread on a screw is a helix.

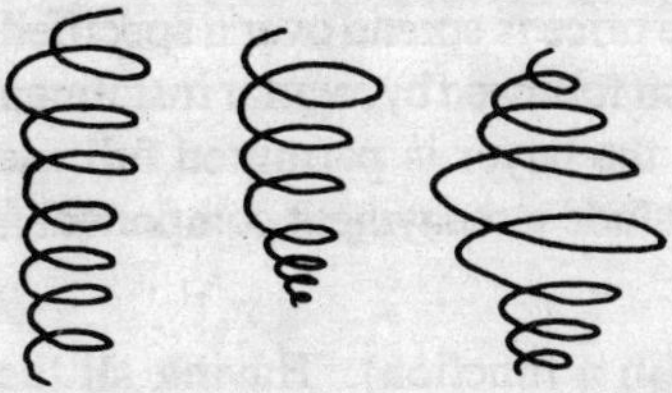

Fig. H-1.

Henry. Symbol: *H.* Refers to the SI unit of inductance which is equal to the inductance of a closed circuit which has a magnetic flux of one weber per ampere of current in the circuit. $1\ H = 1 Wb\ A^{-1}$.

Hero's Formula. Refers to a formula for the area of a triangle with sides a, b and c:

$$A = \sqrt{[s(s-a)(s-b)(s-c)]}$$

where s is half the perimeter.

Heron's (Hero's) Formula. A formula connecting the area of a triangle with its sides. If a, b, c are the length of the sides of a triangle its area Δ is given by

$$\Delta = \sqrt{s(s-a)(s-b)(s-c)}$$

where $s = \frac{a+b+c}{2}$

s is called semi-perimeter.

Hexagon. A plane figure having six straight sides. A *regular hexagon* is one having all six sides and all six angles equal, the angles all being120°. Congruent regular hexagons can be fitted together to cover completely a plane surface. Apart form squares and equilateral triangles, they are the only regular polygons having this property.

Hexahedron. Refers to a polyhedron which has six faces. For example, the cube, the cuboid, and the rhombohedron are all hexahedrons. The cube is a *regular hexahedron;* all six faces are congruent squares.

Hire Purchase. Refers to a method of buying goods wherein payment of the purchase price is spread over a specified time by payment of an initial deposit followed by regular instalments. Once the deposit has been paid the buyer is permitted full use of the goods. All instalments include a repayment component and an interest component.

Homogeneous. (1) (in a function). Having all the terms to the same degree in the variables. For a homogeneous function f (x, y, z, ...) of degree n

$$f(kx, ky, kz, \ldots) = k^n f(x, y, z)$$

for all values of k. For example, $x^2 + xy = y^2$ is a homogeneous function of degree 2 and

$$(kx)^2 = kx.ky + (ky)^2 = k^2 (x^2 + xy + y^2)$$

(2) Describing a substance or object in which the properties do not change with position; for example, the density is constant throughout.

Horse-Power (H.P.). Refers to a unit of power (rate of doing work) equal to 550 foot pounds per second. It can be interpreted as a mechanical horse, having power of 25 lb and moving at the rate of 15 miles per hour that would exert 1 horse power (H.P.).

Hybrid Computer. Refers to a combination of analog and digital computer wherein advantages of both type having been incorporated. It is mainly used for scientific and technological purposes.

Hydrodynamics (or fluid dynamics). The branch of mechanics concerned with the study of ideal fluids in motion. An ideal fluid is assumed to be incompressible and to be free from frictional forces. Although never achieved, this simplification is often necessary to analyse a complex situation. The velocity, acceleration, and pressure at each point in the flow of an ideal liquid gives an indication of what will happen in a real liquid.

Hydrostatics. The branch of mechanics concerned with fluids at rest. It is generally considered to have been instigated by Archimedes. Applications include the design of storage tanks, gates, and valves for hydraulic structures, dams, etc.

Hyperbola. Refers to a conic having eccentricity greater than 1. It has two branches extending upto infinity, two axes of symmetry, two foci, two directrices, two vertices, and two axes, i.e., transverse and conjugate and has a centre that is the midpoint of the line of two foci. Transverse axis is the line that join the foci and perpendicular. Conjugate axis is the line passing through the centre and to the transverse axis. Standard equation of hyperbola is

$$\frac{x^2}{a^2} - \frac{y^2}{b^2} = 1$$

where a and b are lengths of semi transverse and semi conjugate respectively, the centre being at the origin. Coordinate axes are along the transverse and conjugate axes. Vertices are $(\pm a, 0)$, foci $(\pm ae, 0)$ where $e = 1 + \frac{b}{a^2}$ the eccentricity. Equations of the directrices are $x = \pm a/e$. It has two asympotes, $y = \pm\left(\frac{b}{a}\right)x$. If $b = a$ it is a rectangular hyperbola with equation $x^2 - y^2 = a^2$. If the axes are rotated through an angle $\frac{\pi}{4}$ the equation of the rectangular hyperbola will be $xy = c^2$ wherein coordinate axes will be its asympotes.

Hyperbolic Functions. Refers to functions having an analogy to circular functions, called hyperbolic sine, hyperbolic cosine etc. and defined as

$$\text{while } \sinh x = \left(\frac{e^x - e^x}{2}\right), \quad \cosh x = \left(\frac{e^x + e^{-x}}{2}\right)$$

$$\tanh x = \sinh x/\cosh x, \ \coth x = \frac{1}{\tanh x},$$

$$\operatorname{sech} x = \frac{1}{\cosh x}, \quad \operatorname{cosech} x = \frac{1}{\sinh x}.$$

These are related to circular functions by the equations

$$\sin(ix) = i \sinh x, \ \cos(ix) = \cosh x \text{ etc.}$$

Also $\cosh^2 x - \sinh^2 x = 1$, $\operatorname{sech}^2 x + \tanh^2 x = 1$,

$\coth^2 x - \operatorname{cosech}^2 x = 1$,

$\sinh 2x = 2 \sinh x \cosh x$, $\operatorname{coch} 2x = \cosh^2 x + \sinh^2 x$.

Hypocycloid. The locus of a point on the circumference of a circle that rolls on the inside of a fixed circle. Its parametric equation are

$$x = (a - b)\cos\theta + b\cos[(a - b)\,\theta/b],$$

$$y = (a - b)\sin\theta - b\sin[(a - b)\,\theta/b],$$

where a is the radius of the fixed circle, b the radius of the rolling circle and θ the angle between the line of centres and the positive direction of x-axis. There is a cusp of the first kind at every point common to the curve and the fixed circle. The number of cusps is given by $n = a/b$. When $n = 4$ the curve is called astroid whose cartesian equation is given by

$$x^{2/3} + y^{2/3} = a^{2/3}.$$

Hypotenuse. Refers to the side opposite to the right angle in a right-angled triangle. The ratios of the hypotenuse length to the lengths of the other sides are used in trigonometry to define the sine and cosine functions of angles.

I

Implicit Relation. Refers to the relation between two variables indirectly through relations with a third variable is known as parameter. For example, if $x = at^2$, $y = 2$ at, this is an implicit relation between y and x through the parameter t.

Improper Fraction. Refers to a fraction wherein the numerator is greatrer than the denominator. If fraction is composed of algebraic expressions, then the fraction is improper if the degree of the numerator is equal to or greater than that of the denominator.

Impulse (Impulsive Force). It is equal to product of force and the time of its application, expressed as $\int_0^t P\,dt$ that comes out to be P.t. if the force P is constant. Since force is the product of mass and acceleration, or m . $\frac{dv}{dt}$.,

$$\text{then} \quad \int_0^t P.\,dt = \int_0^t m\,\frac{dv}{dt}.\,dt = \int_0^t mdv = mv_2 - mv_1.$$

where v_1 and v_2 are initial and final velocities. As in known, product of mass and velocity is the momentum, hence the impulse, for instance, in an impact, is measured by the change in momentum produced during impact.

Incentre. Refers to the centre of the inscribed circle of a polygon. For example, the incentre of a triangle is the common point of intersection of its angular bisectors.

Incircle (Inscribed Circle). Refers to a circle, if it exists, that touches from insides all the sides of a polygon. Compare *circum-circle* or *Circumscribed circle.*

Inclination of Line in Plane. Refers to the angle, known as angle of inclination, made by the line with some fixed line in the plane, measured anticlockwise form the fixed line that is generally the x-axis in the plane.

Inclination of Two Planes. Refers to the smaller angle between the two planes. It is the angle between their normals.

Inclined Plane. It is a type of machine. Effectively a plane at an angle, that can be used to raise a weight vertically by movement up an incline. Both distance ratio and force ratio depend on the angle of inclination (θ) and equal $1/\sin\theta$. The efficiency can be quite high if friction is low e.g. The screw and the wedge.

Increment. Refers to a small difference in a variable. For example, x might change by an increment Δx from the value x_1 to the value x_2. In calculus, infinitely small increments are used.

Indefinite Integral. Refers to the general integration of a function f(x), of a single variable, x, without specifying the interval of x to which it applies. For example, if $f(x) = x^2$, the indefinite integral

$$\int f(x)\, dx = \int x^2\, dx = x^3/x + C$$

where C is an unknown constant which depends on the interval.

Indeterminate Equation. Refers to an equation which has an infinite number of solutions. For example,

$$x + 2y = 3$$

is indeterminate, as an infinite number of values of x and y will satisfy it. An indeterminate equation in which the variables can take only integral values is known as a *Diophantine equation* and it has an infinite but denumerable set of solutions.

Index. Refers to a number which indicates a characteristic or function in a mathematical expression. For example, in y^4, the exponent, 4, is also called the index. Similarly in $\sqrt[3]{27}$and $\log_{10} x$, the numbers 3 and 10 respectively are known as indices (indexes).

Indirect Proof (reduction and absurdum). Refers to a logical argument wherein a proposition or statement is proved by showing that its negation or denial leads to a contradiction.

Induction. Refers to a method that is used in proving that a statement or proposition P(n) is true for all values of n where n is a natural number. It is done in two steps : (1) P (1) should be shown to be true. (2) It should be proved that if P (k) is considered to be true, P (k + 1) must true. If the two steps are established then P(n) is true for all values of n.

Inequally. A mathematical statement that one quantity is greater or less than another.

$a < (>) b$ means a is less (greater) than b.

$a \leq (\geq) b$ means a is less (greater) than or equal to b.

Infinity. In mathematics, a quanitity larger than any that can be specified. The symbols $+\infty$ and $-\infty$ are read as "plus infinity" and "minus infinity" respectively. They indicate infinitely large positive and negative values. $x \rightarrow +\infty$ means that the value of a variable quantity, x, continues to increase and has no maximum.

Inference. Refers to a process of reaching a conclusion from a set of premises. There are different rules for drawing inferences. First rule is of substitution by which the identity.

$$(4a)^2 - (3b)^2 = (4a - 3b)(4a + 3b) \text{ is obtained.}$$

from the identity $x^2 - y^2 = (x - y)(x + y)$. Then the rule of detachment.

Inference in Statistics. Refers to drawing conclusions about a population based on the random sample such that the probability of a correct inference can be obtained according to different hypotheses concerning the population under study.

Infinite Function. A function f(x) such that

$$\lim_{x \rightarrow a} f(x) = \infty \text{ or } -\infty$$

is infinite function.

Infinite Series. Refers to a series with infinite number of terms.

Infinite Set. Refers to a set wherein the number of elements is infinite. For example, the set of'positive integers', z = {1, 2, 3, 4, ...}, is infinite but the set of 'positive integers less than 20' is a *finite set.* Another example of an infinite set is the number of circles in a particular plane.

Infinity. Symbol : ∞. Refers to the value of a quantity which increases without limit. For example, if $y = 1/x$, then y becomes infinitely large, or approaches infinitely, as x approaches 0. An infinitely large negative quantity is denoted by $-\infty$ and and infinitely large positive quantity by $+\infty$. If x is positive, [$y = -(1/x)$] tends to $-\infty$ as x tends to 0.

Information Theory. The mathematical theory of communication, involving analysis of the information content of messages and the processes used in their transmission, reception, storage and retrieval. Statistical concepts, such as probability, are used to assess the extra information (redundancy) necessary to compensate for spurious signals (noise) occurring during communication. The bit is the basic unit of information and the channel capacity is a measure of the ability of the transmission medium, e.g. a telegraph line, to carry information. Information theory was developed in the 1940s by Claude E. Shannon (1916-). It has been applied widely in such fields as computer science and cybernetics and has aided the understanding of many processes, ranging from the laws of thermodynamics to the use of language.

Initial Conditions. Refers to the value of the vector quantities such as forces, velocities as applicable to a body at the beginning of its movement.

Induction.

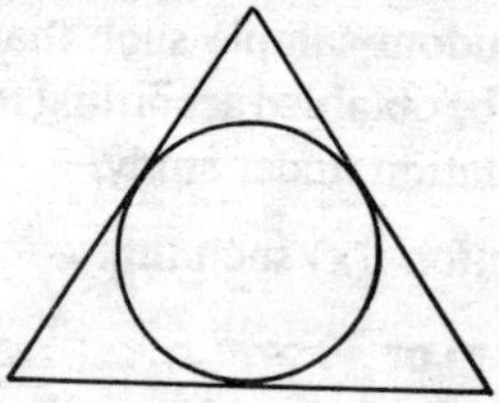

Fig. I-1.

Integers. Symbol: z. Refer to the set of whole numbers {....., $-2, -1, 0, 1, 2, \ldots$} used for counting. It includes zero and the negative whole numbers.

Integer Variable. Refers to a variable that can assume integral values only. In programming, any variable name starting with any one of the letters I, J, K, L, M and N.

Integral Calculus. Refers to that branch of mathematics which deals with integrals of functions.

Integral Curves. Refer to a family of curves whose equation are the solutions of a given differential equation. For example, the family of circles $x^2 + y^2 = c^2$, which is the solution of the differential equation $x + y \frac{dy}{dx} = 0$.

Integral Domain. Refers to a commutative ring having identity element in it with respect to multiplication and with the property that when xy = 0(0 being identity element with respect to addition) then x = 0 or y = 0 both x = y = 0.

Integral of Function. Given a function f(x), let there be a function g(x), whose derivative is f(x) then we say that the integral of f(x) is g(x), written as $\int f(x)dx = g(x)$. We can also write $\int f(x)\,dx = g(x) + c$, where c is an *arbitrary constant of integration* known as constant. This is known as *indefinite integral* of f(x), since no definite value of the integral is known, c being arbitrary. Now, if x lies in between two extreme values a and b, then the *definite integral* of f(x) over the interval (a, b) is the difference between the values of the indefinite integral of f(x) for x = a and x = b, written as

$$\int_b^a f(x)dx = \Big[g(x) + c \Big]_{x=a}^{b} = g(b) — g(a).$$

This represents the *area under the curve* y = f(x) bounded by the curve, the x-axis, and the two ordinates at x = a and x = b.

Integrating Factor. Refers to a multiplier used to simplify and solve differential equations, given the symbol ξ. For example, xdy—ydx = $2x^3dx$ may be multiplied by ξ (x) = $1/x^2$ to give the standard form:

$$d(y/x) = (x.dy — y.dx)/\, x^{32} = 2\,xdx$$

which has = (x. dy — y.dx)/x^2 = 2xdx,

which has the solution $y/x = x^2 + c$,

where c is a constant.

Integration. The process of the continuous summing of change in a function, f(x), over an interval of the variable x. It is the inverse process of differentiation in calculus, and its result is called the *intetgral* of f(x) with respect to x. For example, the total distance a car travels along a road in time interval t_1 to t_2 is the integral of its velocity v over this interval, written

$$\int_{t_1}^{t_2} v.\, dt$$

as this has defined limits t_1 and t_2 it is known as *definite integral*.

In general it is the area between the curve and the x-axis, between the values x_1 and x_2. It can be considered as the sum of a number of column areas of width Δx and heights given by f(x). As Δx approaches zero, the number of columns increases infinitely and the sum of the column areas approaches the area under the curve.

Integration by Parts. Refers to a method of integrating a function of a variable by expressing it in terms of two parts, both of which are differentiable functions of the same variable. A function f(x) is written as the product of u(x) and the derivative dv/dx. The formula for the differential of a product is:

$$d(u.v)/dx = u.dv/dx + v.du/dx$$

Integrating both sides over x and rearranging the equation we get

$$\int u.(dv/dx)\, dx = uv - \int v\, (du/dx)\, dx$$

which can be used to evaluate the integral of a product. For example, in calculating the intetgral of x sin x dx, u is substituted for x and dv/dx for sin x, so that $v = -\cos x + c$, where c is a constant. This gives :

$$x. \sin x\, dx = -x \cos x + \cos x.1.dx = -x \cos x + \sin x + k$$

where k is an arbitrary constant, generally a trigonometric or exponential function is chosen for dv/dx.

Integration by Substitution. Refers to a method of integrating a function of one variable by expressing it as a simpler or more easily integrated function of another variable. For example, to integrate $\sqrt{(1 - x^2)}$ with respect to x, we can make $x = \sin u$, so that $\sqrt{(1-x^2)} = \sqrt{(1 - \sin^2 u)} = \sqrt{(\cos^2 u)} = \cos u$, and $dx = (dx/du)$, $du = \cos u du$. Therefore:

$$\int_a^b \sqrt{(1-x^2)}\,dx = \int_c^d \cos^2 u.\,du$$

$$= \Big[u/2 - \tfrac{1}{2}\sin u \cos u \Big]_c^d$$

In substitution the limits are also changed to corresponding values of u.

Interest. The amount of money charged by a lender to a borrower for the use of a loan. The principal (P) is the amount on which interest is calculated; the term (t), the length of time in years for which the money is lent; and the interest rate(r), the annual rate of return per 100 units of principal. In simple interest, the principal each year is the sum originally lent. The lender is paid Prt/100 in interest and repaid P after t years. In compound interest, the interest each year is not paid to the lender but is added to the principal, so that the principal for the next year becomes P(1+ r/100). After t years the lender is paid $p(1 + r/100)^t$.

Interface. Refers to a shared boundary. It is the area (s) or p!ace (s) where two devices or two systems meet and interact. There is a simple interface between an electric plug and socket. A far more complicated interface of electronic circuits makes the connection between the central processor of a computer and each of its peripheral units. Man-machine interface refers to the interaction betweeen people and machines, including computers. For good, i.e., efficient, interaction devices, like visual display units and easily understandable programming languages have been introduced.

Interpolation. Refers to the process of estimating the value of a function from known values on either side of it. For examole, if the speed of an engine, controlled by a lever, increases from 40 to 50 revolutions per second when the lever is pulled down by four centimetres, one can interpolate from this information and assume that moving it two centimetres will give 45 revoiutions pér second. This is the simplest method of interpolation, called *linear interpolation*. If known values of one variable y are plotted against the other variable x, an estimate of an unknown value of y can be made by drawing a straight line between two nearest known values.

The mathematical formula for linear interpolation is ;

$$y_3 = y_1 + (x_3 - x_1)(y_2 - y_1)/(x_2 - x_1)$$

y_3 is the unknown value of y (at x_3) and y_2 and y_1 (at x_2 and x_1) are the nearest known values, between which the interpolation is made. If the graph of y against x is a smooth curve, and the interval between y_1 and y_2 is small, linear interpolation can give a good approximation to the true value, but it (y_2 - y_1) is large, it is less likely that y will fit sufficiently well to a straight line between y_1 and y_2. A possible source of error occurs when y is known at regular intervals, but oscillates with a period shorter than this interval.

Intersection. (1) The point where two or more lines cross each other, or a set of points which two or more geometrical figures have in common.

(2) In set theory, the set formed by the elements common to two or more sets. For example, if set A is (black fourlegged animals) and set B is (sheep) then the intersection of A and B, written $A \cap B$, is {black sheep}. This can be shown on a Venn diagram by the intersection of two circles, one representing A and the other, B.

Interval. Refers to a set of all the numbers or points lying between two end points. If x = a and x = b are the extreme values that $\{x \mid a < x < b\}$ is known as an *open interval* represented by (a, b) whereas $\{x \mid a \leq x \leq b\}$ is known as closed interval represented by[a, b].

Intrinsic Equation of Curve. Refers to the equation of a curve in terms of either the radius of curvature ρ and s the arc length or in terms of s the arc length and the angle between the tangent and the positive direction of x-axis.

Invariant. Refers to the property of a function, an equation, or a curve that do not change under a particular transformation. For example, in the general equation of a conic

$$ax^2 + 2hxy + by^2 + 2gx + 2fy + c = 0$$

the expressions (a + b) and (h^2 — ab) are invariants under the transformations like translation, or rotation, or both.

Inverse Element. In a set, the inverse of an element a under the binary operation, is the element a^{-1} such that a* a^{-1} a = e where e is the identity element of the set. For example, in the set of integers, —a is the inverse of a since a + (—a) = (—a) + a = 0, 0 being the identity element with respect to addition.

Inverse Function. If a function f maps a set A into a set B then there may exist a reverse mapping denoted by f^{-1} from the set B into the set A. Thus, if y = f(x) for x ∈ A, y ∈ B then x = f^{-1} (y). A function that allows such an inverse function is known as *bijective function* and the mapping as *bijective* mapping.

Inverse of Matrix. A matrix B of the same order as a square matrix A will be known as the inverse of A if B.A = I = A.B, where I is the unit matrix of the order of A. Here B is written as A^{-1}. Also A^{-1} adjoint A/ | A | where | A | = determinant of A.

Inverse Relation. If there is a relation R between two sets A and B so that xRy for X ∈ A, Y ∈ B, then there exists a relation R^{-1} so that y R^{-1} x.

Inverse Square Law. Refers to a law according to which the intensity of energy or a force changes as the square of the distance from the source. For example, Force of attraction between any two objects differs as the square of distance between them.

Inverse Trigonometric (Circular) Functions. If y = sin x, then we can say that x is the angle whose sine is y and write x = $\sin^{-1}$ y. The function $\sin^{-1}$ is known as inverse since function. Similarly, cos—^{1}, $\tan^{-1}$, $\cot^{-1}$, $\sec^{-1}$, cosec^{-1} are inverse functions corresponding to the trigonometric functions cosine, tangent, cotangent, secant, cosecant respectively.

Involute. The involute of a curve is a second curve that would be obtained by unwinding a taut string wrapped around the first curve. The involute is the curve traced out by the end of the string.

Irrational Number. Refers to number that cannot be written as an exact ratio of two integers. For example, the square root of three. ($\sqrt{3} = 1.7320508...$) can be calculated to any desired degree of accuracy, but can only be exactly defined as the largest number of which the square does not exceed three. Such a number is known as a surd. Another type of irrational number is the transcendental number, that arises not as the result of a simple algebraic relationship, but as a basic property of mathematics. For example, π and e are transcendental numbers, which arise in geometry and calculus respectively. Transcendental numbers are defined as numbers that are not the roots of any algebraic equation with rational coefficients.

Isolated Point. Refers to a point that satisfies the equation of a curve but is not on the main arc of the curve. For example, the equation $y^2 (x^2 - 4) = x^4$ has a solution at $x = 0$ and $y = 0$, but there is no real solution at any point near the origin, so the origin is an isolated point.

Isomorphism. Refers to a one-to-one correspondence between two sets. Each element of the first set can be related to an element in the second set by an operation. For example, multiplication by a constant integer relates one set of integers to another set of integers. A relationship that does not conform to this, for example, taking the square root, will form a *non isomorphism*.

Iteration. Refers to a method of solving a problem by successive approximations, each using the preceding approximation as a staring point to obtain a more accurate estimate. For example, the square root of 3 can be calculated by writing the equation $x^2 = 3$in the form $x = 1/2\,(x + 3/x)$. To obtain a solution for x by iteration, we might start with a first estimate, $x_1 = 1.5$. Substituting this in the equation gives the second estimate, $x^2 = 1/2\,(1.5 + 2) = 1.75000$. Continuing in this way, we obtain:

$$x_3 = 1/2\,(1.75 + 3/1.75) = 1.732\ 14$$

$$x_4 = 1/2\,(1.732\ 14 + 3/1.732\ 14) = 1.73\ 205$$

and so on, to any required accuracy. The difficulty in solving equations by iteration is in finding a formula for iteration (algorithm) that gives convergent results. In this case, for example, the algorithm $x_{m+1} = 3/\, x_n$ does not give convergent result. There are several standard techniques, like Newton's method, for obtaining convergent algorithms. Iterative calculations, although often tedious for manual computation, are widely used in electronic digital computers.

J

Job. Refers to a unit of work to be run on a computer. Generally one job includes a few programmes. The necessary instructions to run a job on a particular computer installation form a part of input and are fed in the Job Control Language (JCL) of that installation. The JCL is interpreted by the computer and is made use of to identify the job and specify its requirement to the computer installation.

Jordan Curve. Refers to a continuous, closed curve not intersecting itself. It may or may not have a finite length.

Jordan Curve Theorem. Refers to a jordan curve that divides the plane into two regions having the curve as common boundary.

Joule. Symbol : J. The SI unit of energy and work that is equal to the work done when the point of application of a force of one Newton moves one metre in the direction of action of the force. 1 J = 1 Nm. The joule is the unit of all forms of energy.

K

Kendall's Method. Refers to a method of measuring the degree of association between two different ways of ranking n objects using two variables (x and y), which give data (x_1, y_1), ..., (x_n, y_n). The objects are ranked using first the x's and then the y's. For each of the 2n (n — 1)/2 pairs of objects a score is assigned. If the rank of the jth object is greater is greater (or less) than that of the kth, regardless of whether the x's or y's are used, the score is plus one. If the rank of the jth is less than that of the kth using one variable but greater using the other, the score is minus one. Kendall's coefficient of rank correlation τ = (sum of scores)/1/2 n (n — 1). The closer τ is to one, the greater the degree of association between the rankings.

Kernel, Iterate Kernels (Integral Equations). The functions K_n defined by $K_1(x, y) = K(x, y)$ and

$$K_{n+1}(x, y) = \int_a^b K(x, t)\, K_n(t, y)\, dt$$

(n = 1, 2, ...), where K is a given kernel. It follows that the *resolvent kernel* k (x. t; λ) is equal to (—1). $\sum^{\infty} \lambda^n K_{n+1}(x,t)$.

Keypunch. Refers to the machine used to punch the cards. By pressing the keys on the punching machine, small rectangular holes are punched on the card. Thus the information is fed on the punch cards through keypunch in coded form, which, in turn, is passed to the computer through card reader.

Kinetic Energy. Symbol : T. It is equal to the work that an object can do because of its motion. For an object of mass m moving with velocity v, the kinetic energy is given by $mv^2/2$. This gives the work the object would do in coming to rest. The rotational kinetic energy of an object of moment of inertia I and angular velocity ω is given by $I\omega^2/2$.

Kite. Refers to a plane figure with four sides, and two pairs of adjacent sides equal. Two of the angles in a kite are opposite and equal. Its diagonals cross perpendicularly, one of them (the shorter one) being bisected by the other. The area of a kite is equal to the product of its diagonal lengths. In the special case in which the two diagonals have equal lengths, the kite is a rhombus.

Klein Bottle. Refers to a curved surface with the unique topological property that it has only one surface, no edges and no inside and outside. It can be thought of as formed by taking a length of flexible stretchy tubing, cutting a hole in the side through which one end can fit exactly, passing an end of the tube through this hole, and then joining it to the other end from the inside. Starting at any point on the surface a continuous line can be drawn along it to any other point without crossing and edge.

Fig. K-1. The Klein bottle a single closed surface with no inside.

Konigsberg Bridge Problem. A classical problem in topology. The river in the Prussian city of Konigsberg divided into two branches and was crossed by seven bridges in a certain arrangement. The problem was to show that it is impossible to walk in a continuous path across all the bridges and cross each one only once. The problem was solved by Euler in the eighteenth century, by replacing the arrangement by an equivalent one of lines and vertices. He showed that a network like this (called a *graph*) can be traversed in a single path if and only if there are fewer than three vertices at which an odd number of line segments meet. In this case, there are four.

Kronecker Delta. Refers to the function δ_j^i of two variables i and j defined by $\delta_j^i = 1$ if $i = j$, and $\delta_j^i = 0$ if $i \neq j$. The *generalized Kronecker delta* $\left[\delta^{i_1, i_2, \ldots i_k}_{j_1, j_2, \ldots j_k}\right]$ has k superscripts and k subscripts. If no two superscripts are equal and the subscripts are the same set of numbers as the superscripts, the values is said to be +1 or —1 according as an even or odd permutation is needed to arrange the subscripts in the same order as the superscripts. In all other cases its value is zero.

Kummar's Test for Convergence. Let Σa_n be a series of positive number and $\{p_n\}$ be a sequence of positive numbers. If c_n denotes the number$(a_n / a_{n+1}) p_n — p_{n+1}$, then the series Σa_n converges if there is a positive number 0 and a number N such that $c_n > \delta$ if $n > N$, and the series diverges if $\Sigma 1/p_n$ diverges and there is a number N such that $c_n \leq 0$ if $n > N$.

Kur-tosis (Statistics). A descriptive property of distributions designed to indicate the general form of concentration around the mean. It is often defined by the ratio $B_2 = u_4 / u_2^2$, where u_- and u_4 are the 2nd and 4th moments about the mean. In a normal distribution, $B_2 = 3$. It is **mesokurtic, platykurtic** or **leptokurtic** according as $B_2 = 3$, $B_2 < 3$ or $B_2 > 3$. A platykurtic distribution often appears to be less heavily concentrated about the mean, a leptokurtic distribution to be more heavily concentrated, then the normal distribution.

L

Lamil's Theorem. If there forces act upon a body to keep it in equilibrium the forces must be coplanar and each force will be proportional to the sine of the angle between the lines of action of the other two.

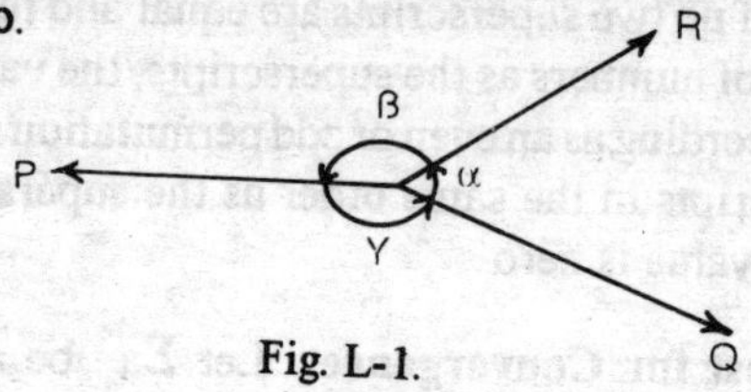

Fig. L-1.

$$\text{Thus } \frac{\vec{P}}{\sin \alpha} = \frac{\vec{Q}}{\sin \beta} = \frac{\vec{R}}{\sin \gamma}.$$

Laplace Transform. The f is the *Laplace* transform of g if

$$f(x) = \int e^{-xt}\, g(t)\, dt,$$

where the path of integration is some curve in the complex plane. It has become customary to restrict the path of integration to the real axis from 0 to $+\infty$. Suppose that g(x) is defined for $x > 0$, has only a finite number of infinite discontinuities, $\int |g(t)|\, dt$ exists for any finite interval, and $f(x) = \int_0^{\infty} e^{-xt} g(t)\, dt$ where this integral converges absolutely for $x > a$. Then this Laplace transformation has an inverse given by

$$g(x) = \frac{1}{2\pi i} \int_{a-i\infty}^{a+i\infty} e^{xt} f(t)\, dt,$$

where the value of the integral is

$$\lim_{a \to 0} \frac{1}{2} [g(x+h) + g(x-h)]$$

If g is of *bounded variation* in the neighbourhood of x and if $\alpha > a$.

Lateral Surface and Area. The **lateral surface** of a surface with bases, such as a cone of cylinder, is the surface with the bases removed. The **lateral area** is the area of the lateral surface.

Latin, Latin Square. An n by n square array which has the property that each row and each column is made up of some permutation of the same n symbol; it is a **diagonal Latin square** if the diagonals also have this property. **In statistics,** such arrays are used to give a method for controlling sources of variability. e.g., suppose one wishes to test five makes of typewriters using five typists and five work-samples. A 5 by 5 Latin square with work-samples as entries describes how each sample can be done by each typists and also be done on each machine, if rows in the square are assigned to typists and columns to machines.

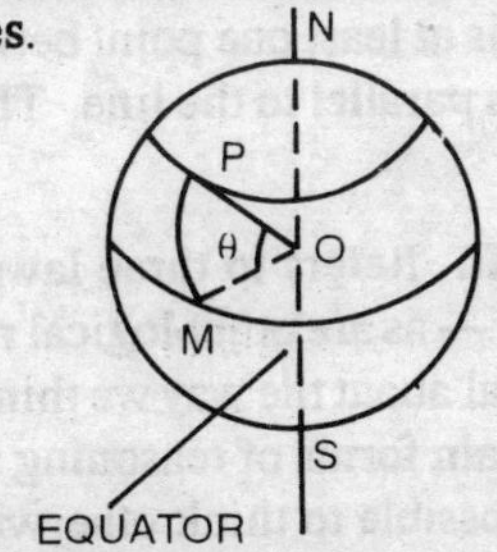

Fig. L-1.

Latitude. Refers to the shortest distance of a point P on the earth's surface from the equator measured as PM along the surface of the earth, as an angle θ subtended at the centre of the earth.

The north pole N and south pole S have latitude 90° and any point on the equator will obviously have 0° latitude.

Latus Rectum. Refers to double ordinate through the focus of a conic section. Half of it is called semi-latus rectum. For example, length of the latus rectum of the parabola $y^2 = 4ax$ is 4a and that of ellipse

$$\frac{x^2}{a^2} + \frac{y^2}{b^2} - 1 \text{ is } \frac{2b^2}{a}.$$

Laws of Impact in Elasticity. If two balls collide while moving in one direction with velocities u_1 and u_2 ($u_1 > u_2$ say) and after the impact their velocities are v_1 and v_2 ($v_1 > v_2$) then,

$$v_1 - v_2 = e(u_1 - u_2).$$

where e is known as the coefficient of restitution which depends upon the elasticity of the materials the balls are made of.

Law to the Mean (mean value theorem). The rule in differential calculus that, if f(x) is continuous in the interval $a \leq x \leq b$, and the derivative f′(x) exists everywhere in this interval, then there is at least one value of x (x_0) between a and b for which

$$[f(b) - f(a)]/(a - b) = f'(x_0)$$

Geometrically this means that if a straight line is drawn between two points, [a, f(a)] and [b, f(b)], on a continuous curve then there is at least one point between these where the tangent to the curve is parallel to the line. This law is obtained from Rolle's theorem.

Laws of Thought. Refers to three laws of logic that are traditionally considered — as are other logical rules — to exemplify something fundamental about the way we think; that is, it is not arbitrary that we say certain forms of reasoning are correct. On the contrary, it would impossible to think otherwise.

(1) *The law of contradiction* (law of non-contradiction). It is not the case that something can both be true and not be true: symbolically

$$\sim(p \wedge \sim p)$$

(2) *The law of excluded middle.* Something must either be true or not be true: symbolically

$$p \vee \sim p$$

(3) *The law of identity.* If something is true, then it is true : symbolically

$$p \rightarrow p$$

Least Squares Method. Refers to a method of fitting a regression line to a set of data. If the data are points (x_1, y_1),... (x_n, y_n), the corresponding points (x_1, y'_1), ...(x_n, y'_n) are found using the line equation y = ax + b. The least squares line minimizes $[(y_1 - y'_1)^2$

$+ / (y_2 - y'_2)^2 + ... + (y_n - y'_n)^2$]. It is found by solving the normal equations.

$$\Sigma y = an + b\Sigma x$$

$$\Sigma xy = a\Sigma x + b\Sigma x^2$$

for a and b. The technique is extended for regression quadra, tics, cubics, etc.

Left-handed Curve. If the torsion of a directed curve C at a point P is positive, then a variable point P moving through the position P in the positive direction along C goes from the positive to the negative side of the osculating plane at P. *Syn.* sinistrorsum [*Latin*], or sinistrorse curve.

Legendre's Equation. Refers to the differential equation of second order of the form

$$(1 - x^2) \frac{d^2y}{dx^2} - 2x \frac{dy}{dx} + (x^2 - n^2)\ y = 0.$$

Leibnitz's Theorem. Refers to a rule for finding the nth derivative of the product of two functions. It is expressed as a series of (n + 1) terms involving successive differentials of separate functions and binomial co-efficients as follows.

$$D^n (uv) = (D^n u)v + n(D^{n-1}\ n)\ (Dv) + \frac{n\ (n-1)}{1.2}(D^{n-2})\ (D^2 v)$$

$$+ \frac{n\ (n-1)\ (n-2)}{1.2.3} (D^{n-3} u)\ D^3 v + ... + u(D^n v)$$

It is quite similar to binomial series for positive integral index.

Lemma. Refers to a theorem sort of thing that has already been proved and is used as a basic assumption in the form of axiom or premise, in another proof.

Leminiscate of Bernoulli. Let O be a point at a distance $a\sqrt{2}$ from the centre C of a fixed circle of radius a and take a variable secant through O which meets the circle in Q and Q'. Now the locus of two points P and P' on OQ such that OP = OP' = QQ' are two branches of the curve known as *Leminiscate* given by Bernoulli. Its cartesian equation is $(x^2 + y^2)^2 = a^2 (x^2 - y^2)$ and the more convenient is the polar equation $r^2 = a^2 \cos 2\theta$.

Lever. Refers to a class of machines or a rigid object able to turn around some fixed point known as fulcrum. These are of, three types:

First Type. In this the fulcrum is between the load and the effort. For example, a crowbar.

Second Type. In this the load is between the fulcrum and the effort. For example, a wheel barrow.

Third Type. In this the effort lies between the fulcrum and the load. For example, a tunning fork.

For a lever of any type to be in equilibrium, the algebraic sum of the three moments about any point must also be zero.

Library Program. Refers to the built-in programs for ready use in a computer. These are used in the form of function subprograms. A good computer installation must have a rich library. For example, library programs for finding the inverse of a matrix, for numerical integration by Simpson's rule, etc.

Light Year. A unit of distance used in astronomy. It is defined as the distance travelled by light in space in one full year. It is approximately equal to 9.467×10^{12} km.

Limacon. In plane it is the locus of a point on a line at a fixed distance from the point of intersection of the line and a fixed circle, as the line rotates about the fixed point on the circle. The equation of limacon is $r = b + 2a \cos \theta$ where a is the radius of the fixed circle and b the fixed distance. If $b < 2a$ there are two loops. If $b > 2a$ there is one loop which approaches to the shape of a circle as a increases. When $b = 2a$ the curve is cardioid with cusp at the pole. The curve was given by Pascal.

Limit. Refers to the value approached by a function as the independent variable approaches some value. The idea of a limit is the basis of the branch of mathematics called *analysis*. There are several examples of the use of limits:

(1) The limit of a function is the value it approaches as the independent variable. For example, the function $x/(x + 3)$ for positive values of x is less than 1. As x increases it approaches

1 — the value approached as x becomes infinite. This is written

$$\lim_{x \to \infty} x/(x+3) = 1$$

stated as 'the limit of x/(x + 3) as x approaches (or tends to) infinity is 1'. 1 is the *limiting value* of the function.

(2) The limit of a convergent sequence is the limit of the nth term as n approaches infinity.

(3) The limit of a convergent series is the limit of the sum of n term as n approaches infinity.

(4) A derivative of a function f(x) is the limit of [f (x + δx) — f(x)]/δx as δx approaches zero.

(5) A definite integral is the limit of a finite sum of terms yδx as δx approaches zero.

Limit of a Function. Strictly, all limits involve functions [see Limit.] Here, we shall be mainly concerned with the limit of a function whose domain and range are sets of numbers. Roughly, a limit is a number that the function approximates closely when a suitable restriction is placed on the independent variable. e.g., the limit of 1/x is 0, as x increases beyond all bounds: the same is true if x takes on alternately, large positive and numerically large negative values, such as 10, —10, —100, 1000, —1000,.... A function is said to *approach its limit or to approach* a certain number as *a limit*. The fact that a function f approaches the limit k as x approaches a given number a is written

$$\lim_{x \to a} f(x) = k$$

and stated "limit of f(x), as x approaches a, is k, *Tech*. A function f has the limit k as x approaches a if, for every positive number ε, there is a number δ such that | f(x) — k | < ε if 0 < | x — a | < δ; f has the limit k as *x becomes infinite* if, for every positive number ε, there is a number δ such that | f(x) — k | < ε if x > δ; f has a limit k as | x | becomes infinite if, for every positive number ε, there is a number δ such that | f(x) — k | < ε if | x| > δ.. Sometimes + ∞ is permitted as a limit; then

$$\lim_{x \to \infty} \text{more } f(x) = +\infty$$

means that for any positive number ε there is a number δ such that $f(x) > \varepsilon$ if $x\ \delta$. There are many other limits of these types e.g., see below, limit on the left or right.

Limit on the Left or Right. The **limit on the right** of a function f at a point a is a number M such that for any $\varepsilon > 0$ there exists a $\delta > 0$ such that $| M - f(x) | < \varepsilon$ if

$$a < x < a + \delta,$$

and the **limit on the left** is a number N such that for any $\varepsilon > 0$ there exists $\delta > 0$ such that $| N - f(x) | < \varepsilon$ if $a - \delta < x < x$. A function is continuous on the right (or left) at a if, and only if the limit on the right (or left) exists and is equal to f(a). Such limits are represented in different ways, e.g., by $\lim_{x \to a} + f(x)$, f(a+), f (a+).

Syn right-hand limit, left-hand limit.

Limit of the Ratio of an Arc to its Chord. Refers to the limit of this ratio when the chord (or arc) approaches zero. If the curve is a *circle*, this limit as 1.

Limit of Sequence. The limit of sequence $\{u_n\}$ is the limit of u_n as n tends to infinity. More precisely, l is the limit of the sequence $\{u_n\}$ written as $\lim_{a \to \infty} u_n = l$, if, given a small positive number ε there exists a number N such that $| l - u_n | < \varepsilon_n$ greater than N.

Line. Refers to a join between two points in space or on a surface. A line has length but no breadth, that is, it has only one dimension. A straight line is the shortest distance between two points on a flat surface.

Linear Equation. Refers to an equation in which the highest power of an unknown variable is one. The general form of a linear equation is

$$mx + c = 0$$

where m and c are constants. On a Cartesian coordinate graph

$$y = mx + c$$

is a straight line that has a gradient m and crosses the y-axis at $y = c$. The equation

$$x + 4y^2 = 4$$

is linear in x but not in y.

Line-al Lineal Element (Differential Equations). A directed line segment through a point, whose slope, together with the coordinates of the point, satisfies a given first-order differential equation.

Linear Combination. A linear combination of two or more quantities is a sum of the quantities, each multiplied by a constant (not all the constants being zero).

Linear Element. For a surface $S : x = x(u, v), y = y(u, v), z = z(u, v)$, and a curve $f(u, v) = 0$ on S, the linear element ds is given by $ds^2 = dx^2 + dy^2 + dz^2 = E\,du^2 + 2F\,du\,dv + G\,dv^2$, where

$$E = \left(\frac{\delta x}{\delta u}\right)^2 + \left(\frac{\delta y}{\delta u}\right)^2 + \left(\frac{\delta z}{\delta u}\right)^2,$$

$$F = \frac{\partial x}{\partial u}\frac{\partial x}{\partial v} + \frac{\partial y}{\partial u}\frac{\partial y}{\partial v} + \frac{\partial z}{\partial u}\frac{\partial z}{\partial v},$$

$$G = \left(\frac{\partial x}{\partial v}\right)^2 + \left(\frac{\partial y}{\partial v}\right)^2 + \left(\frac{\partial z}{\partial v}\right)^2$$

and du and dv satisfy $\frac{\partial f}{\partial u} du + \frac{\partial f}{\partial v} dv = 0$. Also written $ds^2 = g\alpha\beta\, du^\alpha du^\beta$ in tensor notation. The **linear element of a surface** is the element of length ds given by $ds^2 = E du^2 + 2F\,du\,dv + G\,dv^2$, without necessary reference to any particular curve on the surface.

Linear Programming. Refers to the process of finding maximum or minimum values of a linear function under limiting conditions or constraints. For example, the function $x—3y$ might be minimized subject to the constraints that $x + y \leq 10$, $x \leq y$, $x \geq 0$ and $y \geq 0$. The constraints can be shown as the area on a Cartesian coordinate graph bounded by the lines $x + y = 10$; $x = y$, $x = 0$, and $y = 0$. The minimum value for $x — 3y$ is chosen from points within this area. A series of parallel lines $x — 3y = k$ are drawn for different values of k. The line k = — 9 just reaches the constraint area at the point (10, 0). Lower values are outside it, and so x = 10, y = 0 gives the minimum value of x — 3y within the constraints. Linear programming is used to calculate the best possible combination of two or more variable quantities that determine the value of another quantity. In most applications, for example, finding the best combination of quantities of each product from a factory to give the maximum profit, there are many variables and constraints. Linear functions with large numbers of variables and constraints are maximized or minimized by computer techniques that are similar in principle to this graphical technique for two variables.

Line Integral. (contour integral, curvilinear integral). Refers to the integration of a function along a particular path, C, which may be a segment of straight line, a portion of a space curve, or connected segments of several curves. The function is integrated with respect to the position vector $r = xi + yj + zk$, which denotes the position of each point P(x, y, z) on a curve C.

For example, the direction and magnitude of a force vector F acting on a particle may depend on the particle's position in *a* gravitational field or a magnetic field. The work done by the force in moving the particle over a distance dr is F.dr. The total work done in moving the particle along a particular path from point P_1 to point P_2 is the line integral.

$$\text{toman} \int_{P2}^{P1} F.\,dr = F_x\,dx + F_y dy + F_z dz$$

Line Printer. Refers to an output device of a computer system which prints characters (letters, numbers, punctuation marks, etc.) on paper a complete line at a time and can thus operate very rapidly: 1000 lines per minute is a typical speed. The *barrel* (or drum) printer is widely used. It has a set of printing characters embossed around the circumference of a barrel at each character position across the page. The paper is continuous, with a line of perforations between each sheet and sprocket holes along each side to control its movement.

Lissajous Figures. Refers to patterns obtained by combining two simple harmonic motions in different directions. For example, an object moving in a plane such that two components of the motion are at right angles are simple harmonic motions, traces out a Lissajous figure. If the components have the same frequency and amplitude and they are in phase the motion is a straight line. If they are out of phase by 90°, it is a circle. Other phase differences produce ellipses. If the frequencies of the components differ, more complex patterns are produced. Lissajous figures can be demonstrated with an oscilloscope, deflecting the spot with one oscillating signal along one axis and with another signal along the other axis.

Litre Symbol. Refers to a unit of volume defined as 10^{-3} metre^3. It not recommended for precise measurements. Formerly, the litre was defined as the volume of the one kilogram of pure water at 4°C and standard pressure. On this definition, $1l = 1000.028$ cm^3.

Locus. Refers to the path that is traced out by a point constrained to move according to given conditions. The focus may be a point, line, curve, plane, a surface, or a solid. For example, the locus of a point which moves in such a way that its distance from a fixed point is always equal to its distance from a fixed straight line, is a parabola.

Locus of Points. Refers to a set of points, generally defined by an equation that relates coordinates. For example, in rectangular Cartesian coordinates, the locus of points on a line through the origin at 45° to the x-axis and the y-axis is given by the equation $x = y$; the line is said to be the locus of the equation. Any geometrical figure like an interval on a line, a circle in a plane, a cube etc., is a locus of points.

Logarithm. Refers to a number expressed as the exponent of another number. Any number x can be written as $x = z^y$.y is then the logarithm to the base a of x. For example, the logarithm to the base ten of 100 ($\log_{10} 100$) is two, since $100 = 10^2$. Logarithms that have a base of ten are called *common logarithms* (or *Briggsian logarithms*). They are used to do multiplication and division calculation because numbers can be multiplied by adding their logarithms. In general $p \times q$ can be written as $a^c \times ad = a^{(c+d)}$, $p = a^c$ and $q = a^d$. Both logarithms and antilogarithms (the inverse function) are available in the form of printed tables. 0.91×5.12 would be calculated as follows : $\log_{10} 4.91$ is 0.6911 (from tables) and $\log_{10} 5.12$ is 0.7093 (from tables). Therefore, 4.91×5.12 is given by antilog (0.6911 + 0.7093) which is 25.14 (from antilog tables).

Similarly, division can be carried out by subtraction of logarithms and the nth root of a number (x) is the antilogarithm of (log x)/n.

For numbers between 0 and 1 the common logarithm is negative. For example, $\log_{10} 0.01 = -2$. The common logarithm of any positive real number can be written in the form $n = \log_{10}x$, where x is between 1 and 10 and n is an integer. For example,

$$\log_{10} 15 = \log_{10}(10 \times 1.5) = \log_{10}10 + \log_{10} 1.5 = 1 + 0.1761$$

$$\log_{10} 150 = \log_{10}(100 \times 1.5) = -2.1761$$

$$\log_{10} 0.15 = \log_{10}(0.1 \times 1.5)\ 1 = -1 + 0.1761 - 1 + 0.1761$$

is written in the notation 1.1761.

The integer part of the logarithm is known as the *characteristic* and the decimal fraction is the *mantissa. Natural logarithms* (*Napierian logarithms*) use the base e = 2.71828 ...., $\log_e x$ is written as lnx.

Logarithmic Scale. Refers to (1) A line wherein the distance, x, from a reference point in proportional to the logarithm of a number. For example, one unit of length along the line may represent 10, two units 100, three units 1000, and so on. In this case the distance x along the logarithmic scale is given by the equation ; $x = \log_{10} a$. Logarithmic scales form the basis of the slide, rule since two numbers may be multiplied by adding lengths on a logarithmic scale [$\log(a \times b) = \log a + \log b$].

The graph of the curve $y = x^n$, when plotted on graph paper having logarithmic coordinate scales on both axes (log-log graph paper), is a straight line since $\log y = n \log x$. This method can be used to establish the equation of a non-linear curve. Known values of x and y are plotted on log-log graph paper and the gradient n of the resulting line is measured, enabling the equation to be found.

(2) Any scale of measurement that changes logarithmically with the quantity measured. For example, pH is chemistry is a measure of acidity or alkalinity— i.e., of hydrogen-ion concentration. It is defined as $\log_{10}(1[H^+])$. An increase in pH from 5 to 6 represents a decrease in $[H^+]$ from 10^{-5} to 10^{-6}, i.e., a factor of 10. An example of a logarithmic scale in physics is the decibel scale used for noise level.

Logic Gate. Refers to an electronic circuit that carries out logical operations. For examples 'and', 'either—or', 'not', neither—nor', etc. Logic gates operate on high or low input and output voltages. Binary logic circuits, are those that switch between two voltage levels (high and low) and are widely used in digital computers. The *inverter gate* or *NOT gate* simply changes a high input to a low output and vice versa. In its simplest form, the *AND gate* has two inputs and one output. The output is high if and only if both inputs are high. The NAND *gate* (not and) is similar, but has the opposite effect; that is, a low output if and only if both inputs are high. The OR *gate* has a high output if one or more of the inputs are high. The *exclusive* OR *gate* has a high input only if one of the inputs, but not more than one is high. The NOR *gate* has a high output only if all the inputs are low. Logic *gates* are formed using transistors, but in a circuit diagram they are generally shown by symbols which denote only their logical functions. These functions are, in effect, those relationships that can hold between propositions in symbolic logic, and their combinations can be represented in a truth table.

Log-Linear Graph (semilogarithmic graph). Refers to a graph on which one axis has a logarithmic scale and the other has a linear scale. On a log-linear graph, an exponential function (one of the form $y = ke^{ax}$ where k and a are constants) is a straight line. Values of y are plotted on the linear scale and values of x on the logarithmic scale.

Longitude. Refers to the east-west position of a point on the Earh's surface which is measured as the angle in degrees from a standard meridian (taken as the Greenwich meridian). A *meridian* is a great circle which passes through the North and South poles. The *local meridian* of a point is a great circle which passes through that point and the two poles.

Longitudinal Wave. Refers to a wave motion in which the vibrations in the medium are in the same direction as the direction of energy transfer, e.g., Sound waves, transmitted by alternate compression and rarefaction of the medium.

Loop. Refers to a sequence of instructions in a computer program which is performed either a specified number of times or is performed repeatedly until some condition is satisfied.

Lorentz-Fitzgerald Contraction. Refers to a reduction in the length of a body that is moving with a speed v relative to an observer, as compared with the length of an identical object which is at rest relative to the observer. The object is considered to contract by a factor $\sqrt{(1 - v^2 / c^2)}$, c is the speed of light in free space. The contraction was postulated to explain the negative result of the Michelson-Morley experiment making use of the ideas of classical physics. The idea behind it was that the electromagnetic forces holding atoms together were modified by motion through the ether. The idea was made superfluous (along with the concept of the ether) by the theory of relativity, which supplied an alternative explanation of the Michelson-Morley experiment.

M

Machine. Refers to a device used for transmitting force or energy between one place and another. The user applies a force (the effort) to the machine; the machine applies a force (the load) to something. These two forces may not be same ; in fact the purpose of the machine is generally to overcome a large load with a small effort. For any machine this relationship is measured by the *force ratio* (or mechanical advantage) which is equal to the force applied by the machine (load, F_2) divided by the force applied by the user (effort, F_1).

The work done by the machine cannot exceed the work done to the machine. Therefore, for a 100% efficient machine :

if $F_2 > F_1$ then $s_2 < s_1$

and if $F_2 < F_1$ then $s_2 > s_1$.

Here s_2 and s_1 are the distances moved by F_2 and F_1 in a given time.

The relationship between s_1 and s_2 in a given case is measured by the *distance ratio* (or velocity ratio) which is equal to the distance moved by the effort (s_1) divided by the distance moved by the load (s_2).

Neither distance ratio nor farce ratio have a unit ; neither has a standard symbol.

Machine Language (Machine Code). In computer programming all the data and instructions are given to the central processor in a binary code (machine code) having only two digits 0 and 1. Thus, the machine language is all in terms of binary digits only and though it is possible for the programmer to write his program in machine language, it is a very difficult and time consuming process. Therefore, the programs are generally written in high level languages and then automatically translated to machine language by use of compilers.

Maclaurin's Series (Theorem or expansion). If a function f(x) is continuously differentiable any number of times in the neighbourhood of the origin then

$$f(x) = f(0) + xf'(0) + \frac{x^2}{2!} f''(0) + \ldots\ldots$$

Magic Square. Refers to a square array of order n × n developed from the integers 1, 2, 3,.....n where n is the order of the magic square. The magic square is produced in such a way that the sum of the elements in any row, column or diagonal is the same. The three basic squares are given below.

2	9	4
7	5	3
6	1	8

12	7	9	6
13	2	16	3
8	11	5	10
1	14	4	15

9	2	25	18	11
3	21	19	12	10
22	20	13	6	4
16	14	7	5	23
15	8	1	24	27

The element of such arrays can be rearranged also. There in only 1 magic square of orders 3,880 of orders 4 and on unknown number of magic squares of order 5.

Magnetic Disk. Refers to a widely used device for computer systems for storing information. It is a flat circular metallic plate coated on both sides by a magnetised material. On the coated surface there are concentric tracks having spots magnetised in one or two directions so that the information is binary. One disk can store several million characters. Information can be added, altered or deleted as required by magnetic means.

Magnetic Tape. Refers to a long strip of flexible plastic having a magnetic coating on which information can be stored. Its use in the recording and reproduction of sound is well-known. It is also used in computing to store information which is to be fed into a computer and which is obtained from a computer during and after a program has been run. The data is stored on the tape in the form of small closely packed magnetic spots arranged in rows across the tape. The spots are magnetized in one of two directions such that the data is in binary form. The magnetization pattern of a row of spots represents a letter, digit. (0 — 9), or some other character. There are generally nine or seven spot positions across a tape, forming columns or tracks along its length. A number of adjacent rows is used to store an item of information. With 800, 1600, or 6250 rows per inch of tape, one magnetic tape can store a large quantity of information. This information can be altered or deleted as necessary by magnetic means. A tape can therefore be reused many times. The tape must be of very high quality and is typically 1/2 inch (1.27 cm) wide and can be up to 700 metres long. It is generally wound on reels; cassettes are also used.

Information can be recorded on tape making use of a special typewriter ; this method is called *key-to-tape*. The information is fed into a computer making use of a *magnetic tape unit* (MTU). In the simplest version two tape reels are driven at very high speed such that a magnetic tape is would form one reel to the other and back again. As it does so each track on the tape passes close to a small electromagnet, known as *readwrite head*. These extract (read) information required by the central processor of the computer, according to program instructions, and record (write information that is sent from the central processor. A particular piece information can only be read or written when the tape has been wound into position under the heads. The MTU is thus a serial-acess as opposed to a random-acess device. It is widely used as a backing store.

Mass. Refers to the amount of matter contained in a body. The unit of mass is gramme in C. G. S. systems, kilogram in M. K. S. system and pound in F. P. S. system.

Mass-Energy Equation. Refers to the equation $E = mc^2$, where E is the total energy (rest mass energy + kinetic energy + potential energy) of a mass m, c is the speed of light in free space. The equation is a consequence of Einstein's special theory of relativity. The equation is a quantitative expression of the idea that mass is a form of energy and energy also has mass. Conversion of rest-mass into kinetic energy is the source of power in radioactive substances and the basis of nuclear power generation.

Matched, Matched Samples (Statistics). For two random samples,

matched pairs refers to samples paired by consideration of some variable other than the one under immediately study ; e.g., in a study of heights of two groups of ten persons each, the individuals may be paired one from each group so that two persons in a pair have the same age. Similarly, given a set of random samples, a **matched group** or set or **matched samples** refers to a set of samples made up of one sample chosen form each random sample by consideration of some variable other than those under immediate study. The objective is to control variation due to some extraneous factor.

Material, Material Point, Line, or Surface. Refers to a point, line, or surface thought of as having mass. (If one thinks of a lamina with a fixed mass whose thickness approaches zero and density increases proportionally, the limiting situation can be considered as a region with the fixed mass.)

Mathematical, Mathematical System. Refers to one of more sets of undefined objects. Some undefined and defined concepts, and a set of axioms relating to these objects and a concepts. One of the simplest and most important mathematical systems in the *group.* More complicated systems are the real numbers and their axioms (properties) and the plane Euclidean geometry system of points, lines, suitable undefined, and defined concepts like "on a line," "between," "triangle," and "parallel", and axioms concerning these objects and concepts. A mathematical system is an abstract deductive theory that can be applied in other mathematical situations when the axioms can be verified. The success of application in other fields of knowledge depends on how well the mathematical system describes the situation at hand.

Mathematics. The logical study of numerical and spatial relationships. It is usually divided into pure and applied mathematics. In pure mathematics the general theoretical principles are studied, often in abstract. Its branches are arithmetic, algebra, calculus, geometry and trigonometry. Some form of mathematical calculation is an indispensable part of all financial transactions and all measurements. The ancient Egyptians, Sumerians, and Chinese were all using a form of abacus to carry out these calculations for thousands of years before the Christian era. But it was not until the 9th century AD that al-Khwarizmi introduced the idea of writing down calculations instead of carrying them out on an abacus. The Venetian mathematicians of the 11th and 12th centuries were largely responsible for the introduction of these methods to the West; indeed it was they who showed that commercial calculations based on algorisms (a word derived from al-Khwarizmi's name) were superior to those performed on an abacus. However, the application of mathematics to the physical sciences (including astronomy) was largely a 16th-century development inspired by Galileo. It was from this development that applied mathematics grew. It is now largely concerned with mechanics and statistics.

Mathematics of Finance. Refers to the study of the mathematical practices used in brokerage, banking and insurance. *Syn.* mathematics of investment.

Pure mathematics. Refers to the study and development of the principles of mathematics for their own sake and possible future usefulness, rather than for the their immediate usefulness in other fields of science or knowledge. The study of mathematics independently of experience in other scholarly disciplines. Generally the study of problems in applied mathematics leads to new developments in pure mathematics and theories developed as pure mathematics generally find applications later. Thus no sharp line can be drawn between applied and pure mathematics. *Syn.* abstract mathematics.

Mathematical Programming. Refers to the process of finding an optimum value, i.e., a maximum or a minimum value of a function when the variables involved in the function are put to certain conditions in the form of equations and inequalities. The programming is called linear or non-linear according as the function is linear or non-linear respectively.

$$\begin{vmatrix} a & b \\ c & d \\ e & f \end{vmatrix} + \begin{vmatrix} p & q \\ r & s \\ t & u \end{vmatrix} = \begin{vmatrix} a+p & b+q \\ c+r & d+s \\ e+t & f+u \end{vmatrix} \quad \textit{addition}$$

$$k \times \begin{vmatrix} a & b \\ c & d \\ e & f \end{vmatrix} = \begin{vmatrix} ka & kb \\ kc & kd \\ ke & kf \end{vmatrix} \quad \textit{multiplication by a constant}$$

$$\begin{vmatrix} a & b \\ c & d \\ e & f \end{vmatrix} \times \begin{vmatrix} p & q & r \\ s & t & u \end{vmatrix} =$$

matrix multiplication

$$\begin{vmatrix} (ap + bs) & (aq + bt) & (ar + bu) \\ (cp + ds) & (cq + dt) & (cr + du) \\ (ep + fs) & (eq + ft) & (er + fu) \end{vmatrix}$$

the *determinant* of $\begin{vmatrix} a & b \\ c & d \end{vmatrix} = \begin{vmatrix} a & b \\ c & d \end{vmatrix} = ad — bc$

MATRIX: Examples of matrix algebra.

Matrix. A set of numbers, called elements, arranged in rows and columns to form a rectangular array. It is used to assist in the solution of certain mathematical problems. The commutative, associative and distributive laws of matrix arithmetic and algebra are different from those of ordinary arithmetic. The *determinant* of a square matrix is a number, or algebraic expression, that is obtained by multiplication and addition of the elements in a specified way. It has properties that are useful for simplifying and solving sets of simultaneous equations. Single column matrices may represent vectors, enabling them to be handled algebraically and processed by computer.

Maxima, Minima. The necessary condition for a function $y = f(x)$ to have a maxima or minima at $x = a$ is $f\,(2) = 0$. Further, $x = a$ has a maxima or minima if f (a) is negative or positive. More correctly, f(x) in such situations is supposed to have a local maxima or minima at $x = a$.

Mean. Refers to a representative or expected value for a set of numbers. The arithmetic mean or average (called mean) of $x_1, x_2, \ldots\ldots x_n$ given by

$$(x_1 + x_2 + x_3 + \ldots. + x_n)/n$$

If $x_1, x_2, ..., x_k$ occur with frequencies $f_1, f_2, ...f_k$ then the arithmetic mean is

$$(f_1 x_1 + f_2 x_2 + ... + f_k x_k)/(f_1 + f_2 + ... + f_k).$$

When data is classified, as in a frequency table, x_i is replaced by the class mark. The *weighted mean*

$$W = (w_1 x_1 + w_2 x_2 + ... + w_n x_n)/(w_1 + w_2 + ...4w_n)$$

where weight w_i is associated with x_i. The *harmonic mean*

$$H = n/[1/x_1) + (1/x_2) + ... + (1/x_n)]$$

The geometric mean

$$G = (x_1, x_2, ...x_n)^{1/n}$$

The mean of a random variable is its expected value.

Mean Deviation. Refers to a measure of dispersion of elements of a set from its mean or average. If $\bar{x}$ is the mean of $x_1, x_2, ..., x_n$ then the mean deviation of $\bar{x}$ from x is given by

$$\sum_{i=1}^{n} (x_i - \bar{x})/n$$

Mean Value Theorem is Differential Calculus. If a function f(x) is continuous and is continuously differentiable, that is f(x) is differentiable and f(x) is continuous over the closed interval [a, b] then there exists a value c of x in [a, b] such that f(b) — f(a) = (a — b) f' (c). This is known as *Rolle's Theorem*.

Mean Value Theorem in Integral Calculus. If a function f (x) is continuous and single valued over the closed integral [a,b], then there exists a value c of x in [a, b] such that

$$\int_a^b f(x)\,dx = (b - a)\,f(c).$$

Mechanics. Refers to the study of forces and their effect on objects. If the forces on an object or in a system produces no change of momentum the object or system is in equilibrium. The study of such systems is statics. If the forces acting to change momentum the study is known as *dynamics*. The ideas of dynamics relate the forces to the momentum changes produced. *Kinematics* relates to the study of motion without consideration of its cause.

Median. (1) The line joining the vertex of a triangle to the midpoint of the opposite side. (2) The middle value of a set of numbers arranged in order of magnitude. For example, the median of 2, 3, 3, 4, 5 is (3)

Measurable, Measurable Function. A real-valued function f is (Lebesgue) *measurable* if for any real number a the set of all x for which f(x) > a is measurable. A bounded function defined on a set of finite measure is integrable if it is measurable. If g is integrable and $|f(x)| \leq g(x)$ for all x, then f is integrable if it is measurable. Generally, if a measure is defined on an algebra of subsets of a set X then a function having domain X and range continued in a topological space (e.g., the real numbers or the complex numbers) is measurable if the set of all x for which $f(x) \in U$ is measurable for each open set U.

Measure. Means comparison to some unit recognized as a standard.

Angular Measure. Refers to a system of measuring angles.

Board Measure. Refers to a system of measure in which boards one inch or less in thickness are measured in terms of square feet of the side, the thickness being neglected ; those thicker than one inch are measured in terms of the number of square feet on inch thick to which they are equivalent.

Memory, Computer. The part of a computer that stores information. It usually refers to the computer's internal store, in which the programs and data needed to run the computer are held ; this is under the direct control of the central processing unit. In large computers, a high-speed buffer store holds information in active use by the central processing unit, while longer-term storage is provided by the slower main store. Solid -state electronic memory derives operate at very high speeds and are used for the internal memory. Magnetic tapes and disks form the external memory, available to the computer through peripheral devices; it contains information that is not needed frequently.

Meridian. On the *celestial sphere,* a meridian is a great circle that passes through the zenith and the north and south line in the plane of the horizon. On the earth, a meridian is a great circle on the surface of the earth that passes through the geographic poles

Mercator Chart. Refers to a map made by using *Mercator's projection.* A straight line on the plane corresponds to a curve on the sphere cutting meredians at constant angle. Magnification of area on the sphere increases with increasing distance from the equator.

Mercator's Projection. Refers to a correspondence between points of the (x, y) plane and points on the surface of sphere, given by

$$x = k\theta\ ,\ y = k\ \mathrm{sech}^{-1}\ (\sin\phi) = k \log \tan \frac{\phi}{2},$$

where θ is the angle of longitude and ϕ is the angle of colatitude. The correspondence is a conformal one, except for singular points at the poles.

Mersenne Number. Refers to a number of type $M_p = 2^p - 1$, where p is a prime. Mersenne asserted that the only primes for which M_p is a prime are 2, 3, 5, 7, 13, 17, 19, 31, 67, 127, 257. Actually M_{67} and M_{257} are not primes. It is now called M_{61}, M_{89} and M_{107} are primes.

Metric, Metric Density. Let E be a measurable subset of the line (or of n-dimensional Eluclidean space). The metric density of E at a point x is the limit (if it exists) of m $(E \cap I)$ m (I) as the length (for measure) m (I) of I approaches zero, where I is an interval which contains x. The matrix density of E is 1 at all points of E except a set of measure zero ; it is 0 at all points of the complement of E except a set of measure zero.

Metric Space. A set T such that to each pair x, y of its points there is associated a non-negative real number $\rho(x, y)$, known as their *distance,* which satisfies the conditions : (1) $\rho(x,y) = 0$ if, and only if, $x = y$; (2) $\rho(x, y) = \rho(y, x)$; (3) $\rho(x, y) + \rho(y, z) \geq \rho(x, z)$. The function p(x,y) is considered to be a **matrix** for T. The plane and three dimensional space are metric spaces having the usual distance. Hilbert space is also known as a metric space. A topological space is **metrizable** if a distance between points can be defined so the space is a metric space such that open sets in the original space or open sets in the metric space and conversely; i.e., there exists a topological transformation between the given space and a metric space. A compact Hausdorff space is metrizable if and only if it satisfies the *second axiom of countability;* a regular T_1 topological space is metrizable if it satisfies the second axiom of countability (Urysohn's theorem).

Microcomputers. Small sophisticated computers designed for a single user and often for a specific application. The tiny central processing unit is a microprocessor. Microcomputers were introduced in the 1970s for industrial, commercial and domestic uses. They have greatly advanced automation.

Microprocessor. The central processing unit of a microcomputer. Its development was made possible in the 1970s by advances in solid-state electronics, in particular the design of integrated circuits of such complexity that all the main calculating functions can be carried out by a single silicon chip.

Minkowski Distance Function. Relative to a convex body B of which the origin O is an interior point, the Minkowski *distance function* F is defined for each P (other than O) in the space by letting F(P) be the greatest lower bound of the ratio ρ(O, P)/ρ(O, Q). where Q is a point of B on the ray OP and ρ(O, P) represent the distance between O and P. We then define F(O) to be 0 and have F(P) < 1 if P is interior to B, F(P) = 1 if P is on the boundary of B, and F(P) > 1 if P is exterior to B. The function F is a convex function of P. Two **polar reciprocal convex bodies** are any two convex bodies, each containing the origin in its interior, such that the *support function* of each is the *distance function* of the other.

Mil. Refers to a unit of angle measure, equal to $\frac{1}{6400}$ of a complete revolution, 05925°, and nearly $\frac{1}{10000}$ of a radian. Used by U.S. artillery.

Mile. Refers to a unit of linear measure which is equal to 5280 feet or 320 rods.

Geographical Mile. same as Nautical Mile.

Nautical Mile. See Knot, Nautical.

Milnor, John Willard. American algebraic and differential topologist and Fields Medal recipient (1962). Who showed that there are differential manifolds that are homeomorphic to the sevensphere but are not diffeomorphic to it.

Minimal, Adjoint Minimal Surfaces. Two associate surfaces having parameters α_1 and α_2 differing by $\pi/2$.

Associate Minimal Surfaces. When the minimal curves of a minimal surface are parametric, the coordinate functions are of the form $x = x_1(u) + x_2(v)$, $y = y_1(u) + y_2(v)$, $z = z_1(u) + z_2(v)$. The related equations $x = e^{i\alpha} x_1(u) + e^{-i\alpha} x_2(v)$, $y = e^{i\alpha} x_2(v)$, $y = e^{i\alpha} y_1(u) + e^{-\alpha i} y_2(v)$, $z = e^{i\alpha} z_1(u) + e^{-i\alpha} z_2(v)$ define a family of minimal surfaces, known as **associate minimal surfaces** having parameter α.

Double Minimal Surface. Refers to a one-sided minimal surface; a minimal surface S such that, through each of its points P, there exists a closed path C on S with the property that, when a variable point traverses C returning to P, to positive direction of the normal is reversed.

Minimal Curve. A Curve in which the linear element ds vanishes identically. With a Euclidean metric, $ds^2 = dx_1^2 + dx_2^2 + \ldots + dx^2$, this can take place only if the curve reduces to a point of at least one of the coordinate functions is imaginary.

Minimum Point. Refers to a point on the graph of a function where it has the lowest value within an interval. If the function is a smooth continuous curve, the minimum is a turning point, that is, the slope of the tangent to the curve changes continuously from positive and negative by passing through zero. If there is a lower value of the function outside the immediate neighbourhood of the minimum, it is a *local minimum* or *relative minimum*). If it is lower than all other values of the function it is an *absolute minimum*.

Minor in Determinant. The minor of element in a determinant is that determinant of order one less than that of the original determinant, obtained by deleting the row and column there in that element occurs. The minor of an element a_{ij} multiplied by $(-1)^{i+j}$, called *cofactor* of the element a_{ij}.

Minus. It is a latin word for less. It is represented symbolically as when put before any number of an expression it signifies a negative quantity as one 'subtracted' from zero. For example, $-37 = 0 - 37$.

Mobius Band (Strip). Refers to a continuous flat loop having one twist in it. If a flat rectangular strip ACBCD is joined so that D fits B and C fits A by giving the strip a twist in the middle and a Mobious strip is obtained with one continuous surface. Here the peculiarity is that any point on the original strip can be joined to any point on the other side without crossing an edge as it has one edge and one surface only. This is very important in topology.

Modulo (Modular) Arithmetic. The modulo n or mod n arithmetic uses only the numbers 0, 1, 2, ..., n —1. When the fundamental operations are carried out and the number exceeds (n—1) it is divided by n and the remainder is used instead of the earlier number. For instance, the equation $x - 8 = 0$ gives $x = 8$ and if we are having Arithmatic Modulo 7 then $x = 8 \equiv 1 \pmod 7$.

The equation $x + 5 = 0$ gives $x = -5 \equiv 2 \pmod 7$.

Also the equation $2x = 5$ gives $x = 5/2$

but we can write $2x = 5 \equiv 12 \pmod 7$ so that $x = 6$.

Modulus. Refers to the absolute value of a quantity without taking into account its sign. Modulus of x is represented by $|x|$, $|x| = x$ if $x \geq 0$ and $|x| = -x$ if ≤ 0, so that $|5| = 5$ and $[-5] = 5$. The modulus of a vector is its length or magnitude. The modulus of a complex quantity $x + iy$ is equal to $+\sqrt{x^2 + y^2}$.

Moment. In a set of observations of a variable x if d represents deviation (generally from mean) then moment of nth order or nth moment is $\frac{1}{N}\Sigma d^n$ where N is number of observations.

Obviously, the second moment is variance and first moment is zero if the derivations are taken from mean only.

Moment Generating Function (m.g.f.). Refers to a function which is used to calculate the statistical properties of a random variable, x. It is defined in terms of a second variable, t, such that the m.g.f., M (t) is the expectation value of $e^{i\alpha}$, $E(e^{i\alpha})$. For a discrete random variable

$$M(t) = \sum e^{i\alpha} p$$

and for a continuous random variable

$$M(t) = \int e^{ix} f(x)\, dx.$$

Two distributions are the same if their m.g.fs are the same. The mean and variance of a distribution can be calculated by differentiating the m.g.f. The mean E(x) = M' (O) and the variance, Var (x) = M'' (O) — $[M' (O)]^2$.

Moment of Inertia. Symbol : I. Refers to the rotational analogue of mass. The moment of inertia of an object rotating about an axis is given by

$$I = mr^2$$

where m is the mass of an element a distant r form the axis.

Moment of Mass. Refers to the moment of mass of a point mass about a point, line, or plane is the product of the mass and the distance from the point or the mass and the perpendicular distance from the line or plane. For a system of point masses, the moment of masses is the sum of the mass-distance products for the individual masses. For an object the integral must be used over the volume of the object.

Moment of Momentum. The moment of momentum relative to a point O of a particle of mass m moving with velocity v is given by the vector product of the position vectro r of the particle relative to O and the momentum *mv*. In symbols of vector analysis, where a cross denotes the vector product, the moment of momentum is H = r × mv. The moment of momentum relative to O of a system of particles is defined as the sum of the moments of momentum of the individual particles. For a continuous distribution of mass,

$$H = \int_s (r \times v)\, dm,$$

where the integration is extended over the entire body.

Principle of Linear Momentum. Refers to a theorem in mechanics which states that the time rate of change of linear momentum of a system of particles is equal to the vector sum of external forces.

Monodromy, Monodromy Theorem. If the function f of the complex variabie z is analytic at z_o and can be continued analytically along every curve issuing from z_o in a finite simply connected domain D, then f is a function element of any analytic function which is single valued in D; in other words, analytic continuation around any closed curve in D leads to the original functions element.

Monogenic, Monogenic Analythic Function. The totality of all pairs z_o f (z), where

$$f(z) = \Sigma a_n (z - z_o)^n.$$

which can theoretically be obtained directly or indirectly by analytic continuation from a given function element f_o. The function f_o is the **primitive element** of the monogenic function. The Riemann surface of the values z_o is the **domain of existence** of the monogenic function, and the boundary of the domain of existence is the **natural boundary** of the monogenic analytic function. For example, the unit circle $|z| = 1$ is a natural boundary for the function.

$$f(x) = \sum_{n=1}^{\infty} z^{n!}.$$

Monomial Factor. Refers to a single term that may be divided out of every member of an expression ; 3x is a monomial factor of $6x + 9xy + 3x^2$.

Monotonic. Means always changing in the same direction. A *monotonic increasing function* of a variable x increases or stays constant as x increases, but never decreases. A *monotonic decresing function* of x decreases or stays constant as x increases, but never increase. Each term in a *monotonic series* is, in an increasing Monotonic, greater than or equal to the one before it; or, in a decreasing monotonic, less than or equal to the one before it.

Multiple Point. Refers to a point on the curve of a function where several arcs intersect, or which forms an isolated point and at which a simple derivative of the function does not exist. If the equation of the curve is written in the form :

$$(a_1x + b_1y) + (a_2x^2 + b_2xy + c_2y^2) + (a_3x^3 + ...) = 0$$

wherein the multiple point is at the origin of a Cartesian-coordinate system, the values of the coefficients of x and y show the type of multiple point. If a_1 and b_1 are zero, that is if all the first degree terms are zero, then the origin is a singular point. If the terms a_2,b_2 and c_2 are also zero, it is a double point. If, in addition the coefficients a_3, b_3 etc., of the third degree terms are zero, it is a *triple point,* and so on.

Multiplication. Symbol : ×. Refers to the operation of finding the product of two or more quantities. In arithmetic, multiplication of one number, a by another, b, consists of adding a to itself b times. This kind of multiplication is commutative, that is, $a \times b = b \times a$. The identity element for arithmetic multiplication is 1, i.e., multiplication by 1 produces no change. In a series of multiplications, the order in which they are carried out does not change the result. For example, $2 \times (4 \times 5) = 5 \times (2 \times 4)$. This is the associative law for arithmetic multiplication.

Multiplication of vector quantities and matrices does not follow the same rule.

Multiplication of Determinants. Refers to the product of determinants (or of a determinant and a scalar) which is equal to the product of the values of the determinants (or of the value of the determinant and the scalar). The **multiplication of a determinant by a scalar** can be done by multiplying each element of any one row (or any one column) by the scalar. The product of **two determinants of the same order** is another determinant of the same order, in which an element in the ith row and jth column is the sum of the products of the elements in the ith row of the first determinant by the corresponding elements of the jth column of the second determinant (or vice versa). For second-order determinants,

$$\begin{vmatrix} a & b \\ c & d \end{vmatrix} \cdot \begin{vmatrix} A & B \\ C & D \end{vmatrix} \quad \begin{vmatrix} aA + bC & aB + bD \\ cA + dC & cB + dD \end{vmatrix}$$

Multiplication of the Roots of an Equation. Refers to the process of deriving an equation whose roots are each the same multiple of a corresponding root of the given equation. This is effected by the substitution (transformation) $x = x'/k$. The roots of the equation of x' are each k times one of the roots of the equation in x.

Multiplication of Vectors. (1) (*Multiplication by a scalar*). Refers to the product of a scalar a and a vector v is the vector having the same direction as v and of length equal to the product of a and the length of v; i.e., the vector obtained by multiplying each component of v by a. (2) *Scalar multiplication of two vectors.* The scalar product of two vectors is the *scalar* that is the product of the lengths of the vectors and the cosine of the angle between them. This is known as the **dot product**, denoted by **A.B,** or the **inner product.** It is equal to the sum of the products of the corresponding components of the vectors. (3) *Vector multiplication of two vectors.* The vector product of two vectors **A** and **B** is the vector **C** whose length is the product of the lengths of **A** and **B** and the sine of the angle between them (the angle form the first to the second), and which is perpendicular to the plane of the given vectors and directed so that the three vectors in order **A, B, C** form a *positively oriented trihedral.* The product is also called the cross product. If the product of **A** by **B** is **C**, one writes $A \times B = C$. Scalar multiplication is commutative, but vector multiplication is not, for $B \times A = -A \times B$. The scalar product of $2i + 3j + 5k$ and $3j - 4j + 6k$ is $2.3 - 3.4 + 5.6$ or 24, whereas the vector product is $38i + 3j - 17k$ if the vectors are multiplied in the given order, and the negative of this if they are multiplied in the reverse order.

Multiplication Property of one and Zero. For the number one, the property that $a.1 = 1.a = a$ for every number a. For zero, the property that $a.0 = 0.a = 0$ for every number a. The converse of the multiplication property of zero also is true, i.e., if $ab = 0$ then at least one of a or b is zero. This is true for any field (e.g., the field of complex numbers, or arithmetic modulo a prime n) and for any integral domain (e.g., the set of integers), but not for all rings (in arithmetic modulo 6, $2.3 \equiv 0$; and in the ring of 2×2 real matrices,

$$\begin{bmatrix} 1 & 2 \\ 0 & 0 \end{bmatrix} \begin{bmatrix} 0 & 2 \\ 0 & -1 \end{bmatrix} \begin{bmatrix} 0 & 0 \\ 0 & 0 \end{bmatrix}$$

but neither factor in the left member zero matrix)

Multiplier. Refers to (1) The number which is to multiply another number, known as the **multiplication.** (2) In a computing machine, any arithmetic component which performs the operation of multiplication.

N

Napier's Formulae. The formulae given by Napier as are used in spherical trigonometry; these join the sides a, b, c, and opposite angles, α, β, γ of a spherical triangle

$$\sin 1/2\,(a - b)/\sin 1/2\,(a + b) = \tan 1/2\,(\alpha - \beta)/\tan 1/2\,\gamma,$$

$$\cos 1/2\,(a - b)/\cos 1/2\,(a + b) = \tan 1/2\,(\alpha + \beta)/\tan 1/2\,\gamma,$$

$$\sin 1/2\,(\alpha - \beta)/\sin 1/2\,(\alpha + \beta) = \tan 1/2\,(a - b)\,\tan 1/2\,c,$$

$$\cos 1/2\,(\alpha - \beta)/\cos 1/2\,(\alpha + \beta) = \tan 1/2\,(a + b)\,\tan 1/2\,c,$$

Natural Frequency. Refers to the frequency at which an object or system will vibrate freely. A free vibration taking place when there is no external periodic force and little resistance. The amplitude of free vibrations should not be too great. For instance, a pendulum swinging with small swings under its own weight moves at its natural frequency. Normally, an object's natural frequency is its fundamental frequency.

Negation. Symbol : ~ or. In logic refers to the operation of putting *not* or *it is not the case* that in front of a proposition or statement, thereby reversing its truth value. The negation of a proposition *p* is true and vice versa. The truth table definition for negation is illustrated in the illustration.

Negative. Refers to a number of quantity which is less than zero. Negative numbers are also used to denote quantities that are below some specified point. For example, in the centrigrade temperature scale a temperature of —24° C is 24° below the freezing point of water.

Neighbourhood. Means a small interval about a given point. For example, Neighbourhood of 0 is $(-\varepsilon, \varepsilon)$ where ε is a small positive number, thus (—1, 1), (—½, ½), (—1/15, 1/15), (—1/ 100,1/100) etc., are all neighbourhoods of 0. The neighbourhood of the point

$x = a$ is given by $(a - \varepsilon, a + \varepsilon)$.

Nested Intervals. Refers to a sequence of intervals wherein

(i) each interval is included in the preceding interval,

(ii) length of the interval approaches to zero,

(iii) No two different numbers can be in all the intervals of the nest,

(iv) a nest of intervals represents a number if the number lies in side all the intervals of the nest.

Neutral Equilibrium. Refers to equlibrium such that when the system is disturbed a little, there is not tendency for it to move further nor to return.

Newton. A unit of force. A force of one newton is a force which will produce an acceleration of one metre per second to a mass of one kilogram.

Newtonain Mechanics. Refers to mechanics that is based on Newton's laws of motion, i.e., relativistic effects are not taken into consideration.

Newton's Law of Universal Gravitation. The force of gravitational attraction between any two point masses (m_1 and m_2) is proportional to each mass and inversely proportional to the square of the distance (r) between them. Mathematically the law is

$$F = Gm_1 m_2/r^2$$

where G is a constant of proportionality known as the *gravitational constant*. The law can also be applied to bodies; for example, spherical objects can be considered to have their mass acting at the centre.

Newton's Laws of Motion. Three laws of mechanics formulated by Sir Isaac Newton in 1687. They are

(1) An object continues in its state of rest or constant velocity unless acted on by an external force.

(2) The resultant force that is acting on an object is proportional

to the rate of change of momentum of the object, the change of momentum being in the same direction as the force.

(3) If one object exerts a force on another then there is an equal and opposite force (reaction) on the first object exerted by the second.

The first law gives definition of zero force. The second law provides a definition of force based on the inertial property of mass. The third law is equivalent to the law of conservation of linear momentum.

Newton-Raphson Method. Refers to a method used for finding a root of an equation by successive approximations in the form of the iterations. A rough value of root, of course, should be already known for the iterations to start with. The alogrithm or the general formula for successive approximations is

$$x_{n+1} = x_n - f(x_n)/f'(x_n)$$

where x_n is the nth approximation. The formula is obtained by using Taylor's theorem.

Nine Point Circle. In the plane of any triangle ABC there are nine points through which one and only one circle can pass; the nine points are:

Three points are the mid-points D, E, F of the sides BC, CA, AB; three points are the feet of the altitudes G, H, I drawn from the vertices A, B, C; three points are the mid-points J, K, L of the line segements OA, OB, OC of intersection where O is the orthocentre, that is the point of concurrence the latitudes AG, BH, CI. The centre of the circle of these nine points is N which actually is the mid-point of the line segment OT where T is the circumcentre of the triangle ABC. The radius of this circle is equal to $(abc/8\Delta)$ where Δ is the area of the triangle given by A

$$= \sqrt{s(s-a)(s-b)(s-c)} \text{ and } s = (a+b+c)/2.$$

Nominal Value (per value). Refers to the value given to a stock or share by the government or limited company which offers it for sale. Stocks invariably have a nominal value of Rs. 10. Shares, however, may have any nominal value. For example, a company wishing to raise Rs. 100,000 by an issue of shares may issue 100,000 Rs. 1

shares or 200,000 50p shares, or any other combination. The *issue prices*, i.e., the price paid by the first buyers of the shares, may not be the same as the nominal value, although it is likely to be near to it. A share with a nominal value of 50p may be offered at an issue price of 55p ; it is then said to be offered to a premium of 5p. If offered at an issue price of 45p it is said to be offered at a discount of 5p. Once established as a marketable share on a stock exchange, the nominal value has little importance and it is the *market price* at which it is bought and sold. However, the dividend is always expressed as a percentage of the nominal value.

Nomogram. It is a graph consisting of three parallel lines, each one a scale for one of three related variables. A straight line drawn between two points, representing known values of two of the variables, crosses the third line at the corresponding value of the third variable. For example, the lines may shown the temperature, volume, and pressure of a known mass of gas. If the volume and pressure are known, the temperature can be read off the nomogram.

Nomograph (or nomogram). A graph showing the relationship between three variables quantities, enabling the value of one variable to be read off if the other two are known. It can take the form of a series of curves on a graph of two quantities, corresponding to constant values of a third. Or it can consist of three straight

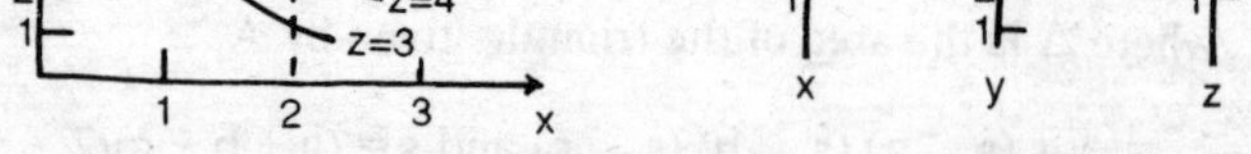

Fig. N-1.

NOMOGRAPH If x = 2 and y =4, a value of approximately 4.5 for z is obtained from the two forms of nomograph shown.

lines calibrated with the values of the variables. A fourth line is drawn between two known points on two of the straight lines : the point at which this fourth line cuts the third straight line gives the value of the unknown quantity.

Non-Euclidian Geometry. Refers to any system of gemoetry wherein the postulates of Euclidian geometry are not valid. In Euclidian geometry we know that from a point outside a given line only one parallel line can be drawn. But in non-Euclidian geometry two possibilities exist :

(i) Through a point outside a straight line or line parallel to the given line can be drawn.

(ii) There are several lines parallel to the given line. When the first possibility is satisfied, it is known as *elliptic geometry* and when the second possibility is satisfied it is known as hyperbolic geometry.

Normal Distribution (Gaussian distribution). The type of statistical distribution followed by, for example, the same measurement taken several times, where the changes of a quantity (x) about its mean value (μ) is completely random. A normal distribution has the probability density function.

$$f(x) = \exp. [-(x-\mu)^2 / 2\sigma^2] / \sigma \sqrt{2\pi}$$

where σ is called the standard deviation. The distribution is written $N(\mu, \sigma^2)$. The graph of f(x) is bell-shaped and symmetrical about $x = \mu$. The standard normal distribution has $\mu = 0$ and $\sigma^2 = 1$. x can be standardized by letting $z = (x - \mu)/\sigma$. The values x^α for which the area under the curve from ∞ to z_α is α, are tabulated, i.e., z is such that $P(z \leq z_\alpha) = \alpha$. Hence $P(a < x \leq b) = P(a-\mu)/\sigma < z \leq (a-\mu)/\sigma$ can be calculated.

Normal Equation of Line. In plane, if p is the length of the perpendicular from the origin upon a line and θ is the angle made by this perpendicular with the positive direction of x-axis then the normal equation of the line is given by $x \cos \theta + y \sin \theta = p$.

Numbers. Mathematical symbols used to denote quantity. The natural numbers, 1, 2, 3, 4, 5...., etc., were developed first by the Hindus and Arabs for simple counting. Substraction led to negative numbers and to zero, which together with natural numbers make up the set of integers ....—3, —2, —1, 0, 1, ... Division of whole numbers results in factions or rational numbers. Some numbers, such as $\sqrt{2}$, cannot be expressed as the ratio of two integers. These are called irrational numbers and they occur as the solutions to simple algebraic equations. They can be calculated to any required accuracy but cannot be written as exact values. Other numbers, called transcendental numbers, do not come from, algebraic relationships. Some of these occurs as basic proporties of space, for example π, the ratio of a circumference of a circle to its diameter. Real numbers include all rational and irrational numbers. The equation $x^2 = -1$ can have no real solution for x, since the square of any number is positive. The imaginary number, $i = \sqrt{-1}$, was introduced to overcome this problem.

Number Sentence. Refers to a statement about numbers that is a sentence if its symbols are interpreted in the usual way. Examples are $3 < 7, 5 - 2 = 3$, $2x + 1 = 17$, $x^2 + 8x = 5$.

Number System. (1) A method of writing numerals to indicate numbers, as for the *binary number system, decimal numbers system,* and *duodecimal number system*. (2) A mathematical system made up of a set of object from known as numbers, a set of axioms, and some operations that act on the numbers, as for the *real numbers system*, the *complex number system,* and *Cayley numbers.*

Number Theory. The study of the properties of numbers. It includes various theorems about prime numbers, many of which are unproved but apparently true, and the study of Diodria), i.e. equations that have only integer solutions. Fermat's last theorem deals with the solution of one of these equations and is a famous unproved theorem. Although number theory has existed for thousands of years, its development has been continuous and it now includes analytic number theory, originated by Euler in 1742; geometric number theory, which uses such geometrical methods of analysis as Cartesian coordinates vectors and matrices;; and probabilistic number theory, based on probability theory.

Numerical Integration. Refers to a procedure for calculating approximate values of integrals. Sometimes a function is called only as a set of values for corresponding values of a variables and not as a general formula that can be integrated. Also, many functions cannot be integrated in terms of known standard integrals. In such cases, numerical integration methods, like the trapezium rule and Simpson's rule, can be used to calculate the area under a graph corresponding to the integral. The area is divided into vertical columns of equal width, the width of each column indicating an interval between two values of x for which f(x) is known. Generally, a calculation is first carried out with a few columns; these are further subdivided until the desired accuracy is attained, i.e., when further subdivision makes no significant difference to the result.

Numerical Phrase. Refers to a collection of numbers and associated signs which indicate how the numbers are to be combined to produce a single number e.g., 3 + 2 (7 — 4) is a numerical phrase which is equal to 9. An **open numerical phrase** means an expression which becomes a number phrase when values are given to certain variables which occur in the expression. e.g., 2x (3 + y) is an open number phrase; when x and y are replaced by 7 and 2, it becomes the numerical phrase 2.7 (3 + 2) which is equal. Refers to 70.

Numerical Sentence. Refers to a statement about numbers e.g., 2 + 3 = 5 is a true numerical sentence and 2 + 5 = 3 is a false numerical sentence. An **open numerical sentence** is an expression which becomes a numerical sentence when values are given to certain variables. e.g., 2x + 3y = 7 is an open numerical sentence. An open numerical sentence is a special type of proposilional function.

O

OCR (optical character recognition). A system used to input information to computer. The information, generally in the form of letters and numbers, is printed, typed, or sometimes hand-written. The characters used can be read by people and can also be read and identified optically by an OCR reader. This machine interprets each character and translates it into a series of electrical pulses. The pulses can then be transmitted to the central processor of the computer.

Octal System. Refers to a number system which is based on the number eight. In the octal number system we have eight digits 0, 1, 2, ...,7 only while in decimal system we have ten digits 0, 1, 2, ...9, and in binary system only two 0,1.

Octant. Refers to one of the eight regions into which space is divided by three mutually perpendicular axes of a three dimensional coordinate system. The first octant has x, y, z, all positive, second octant has x negative, y and z positive and so on.

Odd Function. Refers to a function f(x) of a variable x for which $f(-x) = -f(x)$. For example, sin x and x are odd functions of x.

Oersted. Symbol : Oe. Refers to a unit of magnetic field strength in the c.g.s. system. It is equal to $10^3/4\pi$ amperes per metre ($10^9/4\pi$ Am^{-1}).

Off-Line. Means disconnected from or not under the direct control of a computer. Off-line equipment is either not in use, is undergoing repair, or is doing some task without the assistance of the central processor of the computer.

One-One Correspondence. Refers to a mapping, correspondence, or relation between the elements of two sets such that each element of the first set is mapped into only one element of the second and vice-versa. Symbol used for this is ~ and we say that A ~ B.

On-Line. Means connected to and controlled by a computer. Any device which is connected to and capable of interacting directly with a computer without human intervention is considered to be on-line. In on-line processing the processing of a computer p ogram is done on equipment directly controlled by the central processor.

Open Interval. Refers to a set made up of the numbers between two given numbers (end points), not including the end points, for example, all the real numbers greater than 1 and less than 4.5. The open interval betwen two real numbers a and b is written (a, b). On a number-line, the end points of an open interval are circled.

Open Set. Refers to a set defined by limits which are included in the set itself. The set of all rational numbers greater than 0 and less than ten, writen $(x : 0 < x < 10; x \varepsilon R)$, and the set of all the points inside a circle, but not including the circle itself, are examples, of open sets.

Operating System (OS). Refers to the collection of programs used to control a computer system. It is generally supplied by the computer manufacturer. An operating system has to decide at any instant which of the many demands on the attention of the central processor to satisfy next. These demands include input from and output to different devices, the execution of programs, and accounting and timing. Large computers, wherein many jobs can be run simultaneously, will have a highly complex operating system ; small computers may have a very simple one. A program that runs without the benefit of an operating system is known as *standalone program*.

Operator means. (1) A mathematical function, like addition, substraction, multiplication, or taking a square root or a logarithm, etc.

(2) The symbol denoting a mathematical operation or function, for example : +, —, √, $\log_{10}$.

Operator, Differential. In dy/dx, d/dx ≡ D is a differential operator $D^2 \equiv \left(\frac{d^2}{dx^2}\right)$ is a double differential operator. In $\left(\frac{d^2y}{dx^2}\right) + x\left(\frac{dy}{dx}\right)$ $+4y$, $(D^2 + xD + 4)$ is as differential operator operated on y.

Opposite Angles, Sides, Vertices of Polygon. Refers to angles, sides, vertices that have equal number of sides between them round a polygon.

Opposite Angles, Sides, Vertices of Triangle. In any triangle each side is opposite a vertex or angle, and vice versa.

Order. (1) Refers to (of a matrix). The number of rows and columns in a matrix.

(2) (Of derivative). The number of times a variable is differentiated. For example, dy/dx is a first-order derivative; $\frac{dy^2}{dx^2}$ is second-order; etc.

(3) (Of a differential equation). The order of the higher derivative in an equation. For example,

$$d^3y/dx^3 + 4xd^2y/dx^2 = 0$$

is a third-order differential equation

$$d^2y/dx^2 - 3x\,(dy/dx) = 0$$

is a second-order differential equation.

Order of Root. Refers to the number of times a root of an equation appears.

Order Relation. For a set R of rational numbers, the relation 'V' is an order relation having the following properties (for a, b, c, $\in$ R).

Trichotomy. One of the relations $a < b$, $a = b$, $a > b$ for any a, b, c $\in$ R, is true.

Transitivity. If $a < b$, $b < c$, then $a < c$.

Density. If $a < b$, there is a rational number c such that $a < c < b$.

Extension. For any element a there are two elements b and c such that $b < a < c$.

Consistency. If $a < b$, then $a + c < b + c$ and $ac < bc$ for any $c>0$.

Ordered Pair. Refers to elements which are to be considered only when they are written in a specified order. For example, the coordinates of a point in a plane form and ordered pair (x, y). In general the pair (y, x) will refer to a different point that (x, y).

Ordered Set. Refers to a set of elements that satisfy the following properties. (i) For any two elements a and b, $a < b$, $a = b$, $a > b$ (ii) If $a < b$ and $b < c$ then $a < c$ and if $a > b$ and $b > c$ then $a > c$.

If the above conditions are satisfied only by some of the elements of the set, the set is called a *partially ordered set.*

Ordered Triple. Refers to three numbers indicating values of three variables in a particular order. The x, y and z coordinates of a point in a three-dimensional coordinate system form an ordered triple (x, y, z).

Ordinal Numbers. Refers to whole numbers which denotes order, as distinct from number or quantity. That is, first, second, third and so on.

Ordinary Differential Equation. Refers to an equation which contains total derivatives, but no partial derivatives.

Ordinate. Refers to the vertical or y-coordinate in two-dimensional rectangular Cartesian coordiante system.

Orthocentre. Refers to a point in a triangle it is the point of intersection of lines from each vertex perpendicular to the opposite sides. The triangle obtained by joining the feel of these vertices in the *pedal triangle*.

Orthogonal Projection. Refers to a geometrical transformation which produces an image on a line or plane by perpendicular lines crossing the plane. If a line of length l is projected orthogonally from a plane at angle θ to the image plane, its image length is $l \cos \theta$. The image of a circle is an ellipse.

Orthogonal Trajectory. Refers to a member of family of curves that cut every members of another family of curves orthogonally that is at right angles.

Orthogonal Vectors. Refers to vector quantities which are mutually perpendicular in a plane or space.

Oscillation. Means a regularly repeated movement in one period of oscillation ; or total movement in any vibratory motion in a time taken by particle from one point to the same point returning in the same direction having the same velocity.

Overdraft. Refers to a negative balance on a current amount at a bank. Interest is calculated on overdrafts on a daily basis. An overdraft is distinguished from a bank loan, which is a for fixed sum on all of which interest is payable either monthly or quarterly.

Overflow. Refers to the situation arising in computing when a number, like the result of an arithmetical operation, has a greater magnitude than can be represented in the space allocated to it in a register or a location in store.

Overlay. Refers to a technique that is used in computing when that total storage requirements for a lengthy program exceeds the space available in the main store. The program is split into sections such that only the section or sections required at any one time will be transferred into the main store. These program segments (or overlays) will all occupy the same area of the main store. The overlay structure must be organized such that no routine calls would overwrite it.

P

Pappus's Theorem. Refers to two of the theorems state by Pappus and very useful. First theorem states that when an arc of a plane curve revolves about a line in its plane not cutting the curve, the surface area generated is the product of the length of this arc and the circumference of the circle traversed by its centriod. The second theorem states that the volume of a solid of revolution generated by a plane area that revolves about a line not crossing it, is the product of the area under rotatioin and the circumference of the circle traversed by the centroid of the plane area.

PAR, (Finance). (1) Refers to value stated in a contract to pay, like a bond or not. Also known **par value**. *At par, below par* and *above par refer* to an amount equal to, less than , or greater than the face value (the amount stated in the contract).

(2) The established value of the monetary unit of one country expressed in the monetary unit another; called in full *par of exchange, mint par,* or *commercial par.*

Parabola. The curve formed by a⁺ conic section, in which the distance from a fixed point (focus) and a fixed line (directrix) are equal. In *Cartesian coordinates a standard form of its equation $y^2 - 4ax$, for a parabola that is symmetrical about the x-axis and cuts it at the origin (vertex).

Parabolic Segment. Refers to the area bounded by a chord of a parabola that is perpendicular to its axis and the curve, cut off by the chord. The area of the parabolic segment is $2cd/3$ where c is length of the chord and d its distance from the vertex of the parabola.

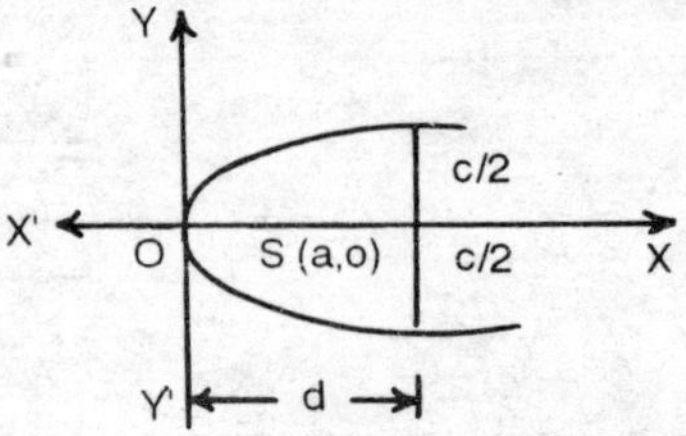

Fig. P-1.

Parabolic Spiral. The spiral wherein the square of the radius vector is proportional to the vectorial angle. Its equation in polar coordinates is $r^2 = a\theta$. Also known as **Fermat's spiral**.

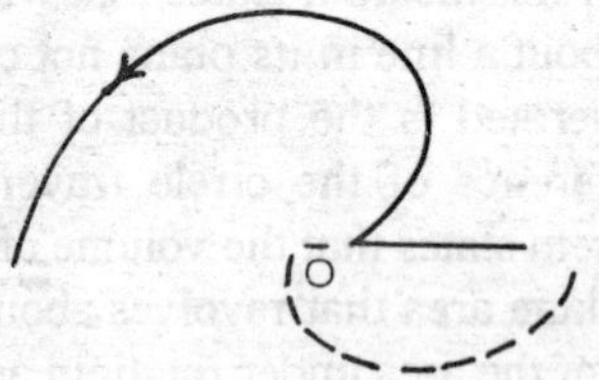

Fig. P-2.

Paraboloid. Refers to the *elliptic* and *hyperbolic paraboloids*. The **elliptic paraboloids** is a surface that can be put in a position such that its sections parallel to one of the coordinate planes are ellipses, and pralled to the other coordinate planes are parabolas. When the surface is in the position illustrated, with its axis along the z-axis, its equation is

$$x^2/a^2 + y^2/b^2 = 2cz.$$

An elliptic paraboloid is a **paraboloid of revolution** if it is produced by revolving a parabola about its axis. This is the special case of the elliptic paraboloid where in the cross sections perpendicular to the axis are circles. The **hyperbolic paraboloid** is a surface that can be put in a position such that its sections parallel to one coordinate plane are hyperbolas, and parallel to the other coordinate planes are parabolas. In the position illustrated, its equation is

$$x^2/a^2 - y^2/b^2 = 2cz.$$

This is a ruled surface, the two families of ruling being

$$x/a - y/b = 1/p, \; x/a + y/b = 2pcz,$$

and

$$x/a + y/b = 1/p, \; x/a - y/b = 2pcz,$$

where p is an arbitrary parameter. These rulings are known as rectilinear generators, because either set may be used to generate the surface.

Paracompact, Paracompact Space. Refers to a topological space T that is a *Housdorff space* and has the property that, for any family F of open sets whose union contains T, there is a *locally finite* family F* of open sets whose union contains T and which are such that each members of F is contained in a member of F. A paracompact space is *regular* and *normal.*

Parallel Curves. Refers to points on the curves with a one-one correspondence such that the tangents at a pair of points are parallel lines, that distance between these parallel lines remain the same for all pairs of points.

Parallel Forces. Refers to forces that act along parallel lines. Resultant of parallel forces can be obtained by adding their magnitude (considering their signs also). Two parallel forces of equal magnitude but in opposite direction and at some distance from each other from a couple and will have a turning effect.

Parallel Planes. Planes lying in space which do not have any common point no matter how far they are extended. For example, equation $2x + 3y - 5z + d = 0$ represents parallel planes for changing values of d.

Parallelogram. Refers to a plane figure haing four straight sides, and with opposite sides parallel and of equal length. The opposite angles of a parallelogram are also equal. Its area is the product of the length of one side and the perpendicular distance form that side to the side opposite. In the some case in which the angles are all right angles the parallelogram is called a rectangle, and when all the sides are equal it is known as a rhombus.

Parallelogram of Vectors. A method for finding the resultant of two vectors acting at a point. The two vectors are represent as two sides of parallelogram: the resultant is the diagonal through the starting point. The method can be used either with careful scale drawing or with trigonometry. The trigonometrical relations give:

$$R = \sqrt{(F_1^2 + F_2^2 + 2F_1F_2 \cos \theta)}$$

$$\alpha = \sin^{-1} [(F_2/F_1) \sin \theta]$$

where θ is the angle between F_1 and F_2 and α is the angle between P and F_1.

Parallelopiped. Refers to solid figure having six faces that are parallelograms. In a *rectangular parallelopiped* the faces are rectangles. If the faces area squares, the parallelopiped is known as a cube.

Hilbert Parallelotope. The set of all points $x = (x_1, x_2, ...)$ of Hilbert space for which $| x_n | \leq (1/2)^n$ for each n. Any compact metric space is homeomorphic to a subset of the Hilbert of parallelotpe. The Hilbert parallelotope is homeomorphic to the set of all x for which $| x_n | \leq 1/n$ for each n, this set is also called a Hilbert parallelotope. Each of these sets is sometimes known as a **Hilbert cube**.

Parameter. Refers to an arbitrary quantity included in an equation that gives a family of curves. For example, in the equation y = mx +c where m is constant and c is arbitrary, c is taken as parameter as this represents a set of parallel lines for different values of c.

Parametric Equation. Here the certesian coordinates x and y are expressed in term of one or two parameters. For example, the parametric equations of the circles $x^2 + y^2 = a^2$ are $x = a \cos \theta$, $y = a \sin \theta$; of parabola $y^2 = 4ax$ these are $x = at^2$, $y = 2at$; of ellipse $x^2/a^2 + y^2/b^2 = 1$ are $x = a \cos \theta$, $y = b \sin \theta$; of hyperbola $x^2/a^2 - y^2/b^2 = 1$, $x = a \sec \theta$, $y = b \tan \theta$, of rectangular hyperbola $xy = c^2$, $x = ct$, $y = c/t$.

Partial Derivative. Refers to the rate of changes of a function of several variables as one of the variables changes and the others are kept constant. For example, if $z = f(x, y)$ the partial derivative $\partial z/\partial x$ is the rate of change of z with respect to x, keeping y constant. Its value will depend on the constant value of y chosen. In the three-dimensional Cartesian coordinates, $\partial z/\partial x$ is the gradient of line at a tangent to the curved surface $f(x, y)$ and parallel to the x-axis.

Partial Differential. Refers to the infinitesimal change in a function of two or more variables resulting from changing only one of the variables while keeping the others constant. The sum of all the partial differentials is the total differential.

Partial Differential Equation. Refers to an equation which contains partial derivatives of a function with respect to a number variables. General methods of solution are available only for certain types of linear partial differential equation that occur in physical problems. *Laplace's equation,* for example, is $(\partial^2 \phi/\partial x^2 + \partial^2 \phi/\partial y^2 + \partial^2\phi/\partial z^2 = 0$ and occurs in the study of gravitational and electromagnetic fields.

Partial Fractions. Refers to a sum of fractions which is equal to a particular fraction, for exanple, $1/2 + 1/4 = 3/5$. Writing a ratio in terms of partial fractions can be helpful insolving equations or calculating integrals. For example

$$1/x\,(x^2 + 1)$$

can be written as

$$a/x + (bx + c)/(x^2 + 1)$$

The values $a = 1$, $b = 1$ and $c = 0$ are then calculated by comparing the coefficients of identical powers of x, to give

$$1/x(x^2 + 1) = 1/x - 1/(x^2 + 1)$$

a form that can be integrated with respect to x as a sum of two integrals.

Particular, Particular Solution (or Integral) of a Differential Equation. Refers to any solution that does not involve arbitrary constants (constants of integration); a solution obtainable from the general solution by substituting special values to the constants of integration.

Partition, Partition of an Integer. The *number of partitions* p(n) of a positive integer n is the number of ways in which n can be written as a sum of positive integers, $n = a_1 + a_2 + ... + a_k$, where k is any positive integer and $a_1 \geq a_2 \geq .... \geq a_k$. If k is restricted so that k + s, than is known as the *number of partitions of n into at most s parts.* Many other types of partitions have been studied. e.g., the number of partitions of n for which the *summands are all different* can be shown to be equal to the number of partitions of n for which the *summands are all odd* (but repetitions are allowed); 5 is equal to 5, 4 + 1, 3+2; it is also equal to 5, 3 + 1+1, 1 +1 + 1 + 1 + 1.

Partition of an Interval. Refers to a set closed intervals $[x_1, x_2]$ $[x_2, x_3]$, ...$[x_n, x_{n+1}]$ for which x_1 and x_{n+1} are the end points of the interval and either $x_i — x_{i+1}$ for all i, or $x_i \geq x_{i+1}$ for all i. The **fineness** (or **mesh** or **norm**) or the partition is the greatest of $[x_{i+1} — x_i]$ for i = 1, 2, ..., n.

Partition of Set. Refers to a finite collection of disjoint sets whose union is the given set. If S has measure (or area, volume etc.), then it may be required that each member of the partition have measure (or area, volume, etc.). Sometimes the requirement the sets be disjoint is changed by the requirement that $(A \cap B)$ has zero measure (or area, volume, etc.) if A and B are members of the partition. The **fineness** of a partition means the least upper bound of distances between points p and q with p and q in the same member of the partition. For a partition p of a set S, any ordered sequence $(A_1, A_2,....)$ whose members are the members of p is an **ordered partition** of S.

Pascal's Distribution (negative binomial distribution). Refers to the distribution of the number of independent Bernoulli trials performed up to and including the rth success. The probability that the number of trials x, is equal to k is given by P (x = k) = $^{k-1}Cr_{-1}\ p^r q^{k-r}$. The mean and variance are r/p and rq/p^2 respectively.

Pascal's Triangle. In accordance with Binomial theorem.

$$(1+x)^n = 1 + {}^nC_1x + {}^nC_2x^2 + {}^nC_3x^3 + ... + {}^nC_rx^r + ... + x^n.$$

If the coefficients are calculated for different values of *n* and arranged in the following pattern, this is called Pascal's triangle:

n = 0							1						
n = 1						1		1					
n = 2					1		2		1				
n = 3				1		3		3		1			
n = 4			1		4		6		4		1		
n = 5		1		5		10		10		5		1	
n = 6	1		6		15		20		15		6		1

and so on.

Payoff. Refers to the amount received by one of the players in a play of game. For a two-person zero-sum game, the **payoff function M** is the function for which M(x, y) (positive or negative) is the sum paid by the *minimizing player* to the *maximizing player* in case the maximizing player uses pure strategy x and the minimizing player uses pure strategy y. A **payoff matrix** for a finite two-person zero-sum game is a matrix so that the element ab_{ij} in the ith row and jth column represents the amount (positive or negitive) paid by the *minimizing player to the maximizing player* in case the maximizing player uses his ith pure strategy and the minimizing player uses his jth pure strategy.

Peano Space. Refers to a space (or set) S that is a *Hausdorff topological space* and is the image of the *closed unit interval* [0,1] for a continuous mapping. A Hausdorff topological space is a Peano space if and only if it is *compact, connected, locally connected, metrizable and nonempty.* A Peano space is also called a **Peano Curve,** although a peano curve generally is to be a curve, [i.e, to have parametric equations of the form $x = f(t)$, $y = g(t)$, where f and g are continuous and $0 \leq t \leq 1$] that passes through each point of the unit square.

Peaison's Classification of Distribution. The differential equation

$$\frac{dv}{dx} = \frac{x + a}{b + cx + dx^2}\, y$$

is satisfied by many common *probability density functions* (e.g., beta, normal, x^2 t. If a probability density function can satisfy an equation of this type, then the values of a, b, c, d and f itself are determined by the first four moments. Pearson classified the probability density functions that are solutions of such a differential equation in accordance with the nature of the zeros of $b + cx + dx^2$. If $a = -\mu$, $b = -\sigma^2$ and $c = d = 0$, then the *normal distribution* with mean μ and variance σ^2 is obtained.

Pedal Curve. At a point P on a curve, tangent PT is drawn where T is the foot of the perpendicular from a fixed point O (generally origin) on the tangent. Then the locus of this point T is called pedal curve. For example, the pedal curve of parabola with respect to focus as the fixed point, is the tangent at the vertex.

Pedal Equation. At a point P on a curve, tangent is drawn. On this tangent, T is the foot of the perpendicular drawn from the pole and p is the length of the perpendicular. The radius vector of P = OP = r. Then a relation between p and r is known as pedal equation of the curve. For example, the pedal equation of the ellipse $x^2/a^2 + y^2/b^2 = 1$, focus taken as pole, is $b^2/p^2 = 2a/r - 1$.

Pedal Triangle. Refers to the triangle that is formed within a given triangle by joining the feet of the perpendiculars form any given point to the sides. The triangle DEF is the pedal triangle formed within the triangle ABC by joining the feet of the altitudes. The figures shows the fact that the altitudes of the given triangle bisect the angles of this pedal triangle.

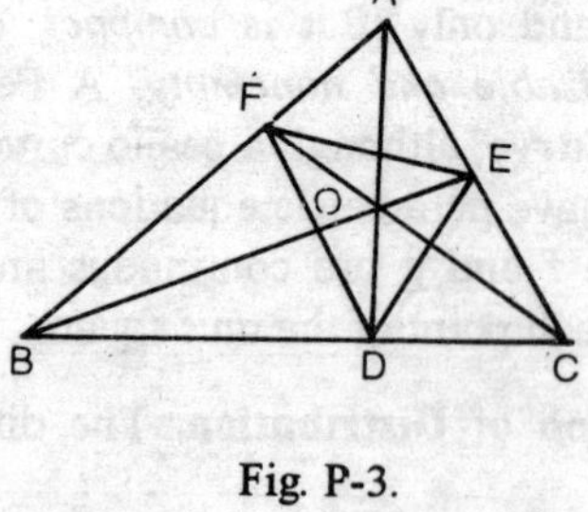

Fig. P-3.

Pencil. Refers to a collection of geometric objects (e.g., a collection of lines or collection of spheres) for which pairs of objects have a common property, e.g., that of intersection of any two of the objects may be the same as that of any other two. In this case, if two different members of the collection have equations f(x, y) = 0 and g (x, y) = 0, then equation of members of the pencil are given by

$$hf(x, y) + kg(x, y) = 0,$$

where h and k are parameters that are not both zero; sometimes one parameter is taken to be 1, but this excludes one member of the pencil [if k = 1, for example, there is no value of h that will reduce the equation of the pencil to f (x, y) = 0].

Pendulum. Refers to body that oscillates freely under gravity. A *simple pendulum* consists of a small mass oscillating to and fro at one end of a weightless cord whose other end is attached to the point of suspension. For small oscillations the motion of the mass is approximately *simple harmonic motion* with *period* of oscillation equal to $2\pi\sqrt{l/g}$ where *l* is the length of the chord from the point of suspension to the centroid of the mass. A *compound pendulum* is a rigid body that oscillates about a point. The period of oscillation depends upon the moment of inertia of the body. For small oscillations the period is the same as that of simple pendulum except that I is to be replaced by $1 + \frac{k^2}{h^2}$ where k is the radius of gyration and *h* is the distance between the point and the centre of mass.

Percentage Error. Refers to the error or uncertainty in a measurement that is expressed as a percentage of the total measurement. For example, if, in measuring a length of 20 metres tape can measure to the nearest four centimetres, the measurement is written as 20 ± 0.04 metres and the percentage error is $(0.04/20) \times 100 = 0.2\ \%$.

Percentile. Refers to one of the set of points that divide a set of data arranged in numerical order into 100 parts. The rth percentile, Pr, is the value below and including which r% of the data lies and above which (100 — r)% lies. Pr can be obtained from the cumulative frequency graph.

Perfect Power. A number or polynomial that is the exact *n*th power of some number of polynomial for a positive integer n>1. **A perfect square** is the exact square of another number of polynomial; e.g., 4 is a perfect square, as is also $a^2 + 2ab + b^2$, which equals $(a + b)^2$. A polynomial like $a^2 + 2ab + b^2$ which is the square of a binomial is a **perfect trinomial square. A perfect cube** is a number or polynomial that is an exact third power, such as 8, 27, and $a^3 + 3a^2b + 3ab^2 + b^3$, which is equal to $(a + b)^3$.

Periodic Motion. Refers to any kind of regularly repeated motion; like the swinging of a pendulum, the orbiting of a satellite, the vibration of a source of sound, or an electromagnetic wave. If the motion can be shown as a pure sine wave, it is known as a harmonic motion. Harmonic motions in general are given by the sum of two more pure sine waves.

Period of Investment. Refers to the length of time for which a fixed amount of capital remains invested. In times of low interest rates if an investor is prepared to commit his money for a long period, as five or ten years, it will gain a higher rate of interest than he can expect on a short-term investment. However, if interest rates are high this will not be the case and long-term rates may be lower than short-term rates.

Period of a Member of a Group. Refers to the least power of the member which is the identity. Also called the **order**. In the group whose members are the roots of $x^6 = 1$ with multiplication as the group operation, $-\frac{1}{2} + \frac{1}{2} i\sqrt{3}$ is of period 3, since

$$\left(-\frac{1}{2} + \frac{1}{2} i\sqrt{3}\right)^2 = 1$$

$$\text{and } \left(-\frac{1}{2} + \frac{1}{2} i\sqrt{3}\right)^2 \neq 1$$

Period Region. For a periodic function of a complex variable, a *period region* is a *primitive period strip* or a *primitive period parallelogram*, according as the function is simply periodic or doubly periodic.

Primitive Period Pair. Two periods ω and ω' of a *doubly periodic function* such that all periods of the function are of the form $n\omega + n'\omega'$, where n and n' are integers.

Primitive Period Strip. If f is a simply periodic function of the complex variable z in the domain D, and ω is a primitive period, then a region in D bounded by line C (or a suitable simple curve extending across D) together with an image of C translated by an amount ω is a **primitive period strip** for f. *Syn.* fundamental period strip.

Periodic, Almost Periodic, Function. A continuous function f is **(uniformly) almost periodic** if, for any $\varepsilon > 0$, the set of all t for which the inequality

$$|f(x+t) - f(x)| < \varepsilon$$

is satisfied for all x has the property that there is a number M so that any interval of length M contains at least one such t, e.g., the function.

Peripheral Unit (peripheral). Refers to a device joined to and controlled by the central processor of a computer. These include input devices, output devices, and backing store. Other examples are visual display units, line printers, magnetic tape units and disk units.

Permissible, Permissible Values of Variable. The values for which a function is defind and that lie on the interval or set on which the function is being considered. Zero is not a *permissible* value for x in the function log x, and 4 is not a permissible value for x if log x is being considered on the interval (1,2). *Permissible* is also used of any values in the domain of the function.

Permutation. Refers to number of possible arrangements in choosing r things from a set of n different things. It is written as ^nPr where P stands for the word permutation.

$$^nPr = \frac{n!}{(n-r)!}$$

Perpendicular Curves. Refers to curves which intersect in such a way that the tangents to two curves at the point of intersection are perpendicular to each other.

Perpendicular Line to Plane. Refers a line that is right angles to any two lines in the plane. It is also called normal to the plane.

Phase Difference. If a simple wave can be represented by the equation

$$y = a \sin (ft - x/\lambda)\, 2\pi$$

the phase of this wave will be $2\pi\ (ft - x/\lambda)$ and the phase difference between two points $x = x_1$, $x = x_2$ at the same time t is equal to $2\pi\ (x_1 - x_2)/\ \lambda$.

Pi. In mathematics, the ratio of the circumference of a circle to its diameter, denoted by the Greek letter π. It was proved to be an irrational number by J. H. *Lambert and transcendental by F. Lindemann (1852-1939). Its value is 3.14159....

Pictogram (pictograph). Refers to a diagram that shows statistical data in a pictorial form. For example, the number of pink, red yellow, and white flowers that grow from a packet of mixed seeds can be shown by rows of the appropriate number of coloured flower shapes.

Pie Chart. Refers to a diagram wherein proportions are shown as sectors of a circles. The relative area of the circle showing the

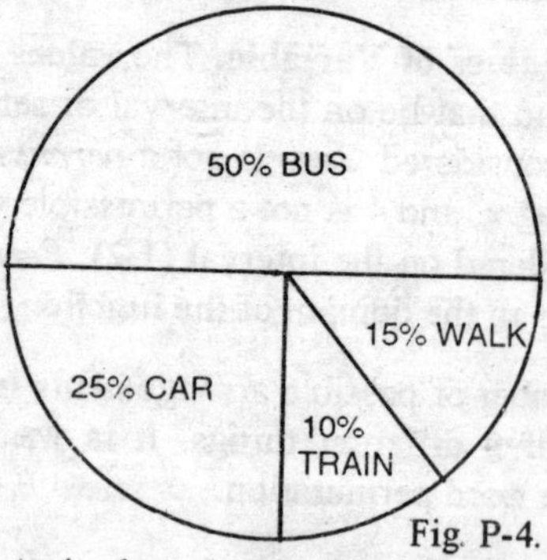

Fig. P-4.
A pie chart showing how a group of workers travel to work.

different proportions. For example, if, out of 100 workers in a factory, 25 travel to work by car, 50 by bus, 10 by train, and the rest walk, the bus passengers are represented by half of the circle, the car passengers by the quarter, the train users by a 36° sector, and so on.

Pint. A unit of capacity. In the UK it is equal to one eighth of UK gallon and is equivalent to 5.6826×10^{-4} m^3. The US liquid pint equal to one eighth of US gallon and is equivalent to 4.7318×10^{-4} m^3. The US dry pint is equal to one sixty-fourth of a US bushel and is equivalent to 5.5061×10^{-4} m^3.

Planar, Planar Point of a Surface. Refers to a point of the surface which $D = D' = D'' = 0$. At a planar point, every direction on the surface is an *asypmtotic direction.* A surface is a plane if, and only if all its points are planar points.

Plane. Refers to a flat surface that contains wholly the straight line drawn through any two points taken on the surface. Its general equation is given by $ax + by + cz + d = 0$ where a, b, c are the direction ratios of the normal to the plane.

Plot. To draw on a graph. A series of individual points plotted on a graph may show a general relationship between the variables represented by the horizontal and vertical axes. For example, in a scientific experiment one quantity can be represented by x and another by y. The values of y at different values of x are then plotted as a series of points on a graph. If these fall on a line or curve, then the line or curve, drawn through the points is a plot of y against x.

Poltter. Refers to an output device of a computer system which produces a permanent record of the results of some program by drawing lines on paper. One pen, or may be two or more pens having different coloured ink, are moved over the paper in accordance with instructions sent from the computer, or from a backing store. Plotters are used for drawing graphs contour maps, etc.

Point of Contact. Refers to a single point where two curves, or two curved surfaces meet. There is only one point of contact between the circumference of a circle and tangent to the circle. Two spheres also do can have only one point of contact.

Point of Inflection. Refers to a point on a curved line where the tangent changes its direction of rotation. Approaching from one side of the point of inflection, the slope of the tangent to the curve increases; and moving away from it on the other side, it decreases. For example, the graph of $y = x^3 - 3x^2$ in rectangular Cartesian coordinates, has a point of inflection at the point $x = 1$, $y = -2$. The second derivative d^2y/dx^2 on the graph of a function $y = f(x)$ is zero and changes its sign at a point of inflection. Thus, in the example above, $d^2y/dx^2 = 6x - 6$, which is equal to zero at the point $x = 1$.

Poisson's Distribution. Refers to a probability distribution of a discrete random variable x. For a variable (r) that can take values in the range 0, 1, 2,..... and has mean value as $P(r) = {}^{r}/r!\ e^{-\mu}$. A binomial distribution with a small frequency of success in a large number n of trials can be approximated by a Poisson's distribution with mean np.

Polar Coordinates in the Plane. The system of coordinates wherein a point is located by its distance from a fixed point and the angle that the line from this point to the given point forms with a fixed line, known as the **polar axis**. The fixed point, O in the figure, is the **pole**; the distance, **OP** = r, from the pole to the given point, is the **radius vector,** the angle θ (taken positive when counter-clockwise) is the **polar angle or vectorial angle**. The polar coordinates of the point P are written (r, θ). The polar angle is also called the **amplitude, anomaly** or **azimuth** of the point. From the figure it can be seen that the relations between rectangular Cartesian and polar coordinates are $x = r\cos\theta$, $y = r\sin\theta$. If r is positive (as in the figure) the amplitude, θ, of the point P is any angle (positive or negative) with OX as initial side and OP as terminal side. If r is negative, θ is any angle having OX as initial side and the extension PO through O as terminal side. The point whose coordinates are (1, 130°) or (—1,—50°) is in the second quadrant; the point whose coordinates are (—1, 130°) or (1, 50°) is in the 4th quadrant.

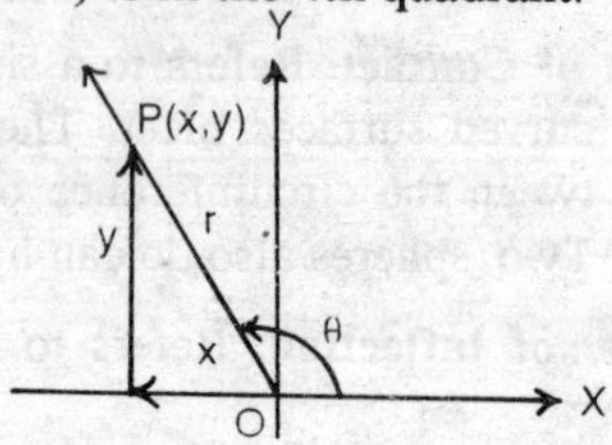

Fig. P-5

Polar form of Complex Number. Refers to the form a complex number takes when it is expressed in polar coordinates.This form is r (cos θ+i sin θ), where r and θ are polar coordinates of the point represented by the complex number. The number r is the **modulus** and the angle θ the **amplitude, argument, or phase** *Syn*. trigonometric form (representation) of a complex number.

Polar Tangent. Refers to the segment of the tangent line to a curve cut off by the point of tangency and a line through the pole perpendicular to the radius vector. The projection of the polar tangent on this perpendicular is known as the **polar subtangent**. The segment of the normal between the point on the curve and this perpendicular is known as the **polar normal.** The projection of the polar normal on this perpendicular is known as the **polar subnormal.**

Polar Triangle of a Spherical Triangle. The spherical triangle whose vertices are poles of the sides of the given triangle. The poles are the ones nearest to the vertices opposite the sides of which they are poles.

Pole and Polar of Conic. If a variable secant is drawn from a given point P to conic meeting the conic in points A and B. On AB find a point Q which is such that P and Q are harmonic conjugates of A and B, that is,

$$PA/PB = QA/QB,$$

then the locus of Q is known as the **polar** of P and P is known as pole. This polar actually is the chord of contact CD of the point P. On CD there would be a point T such that AB is the chord of contact of T.

Polygon. A plane straight-sided geometrical figure with three or more sides; for example, a pentagon, which has five sides. A regular polygon has all sides equal and all interior angles equal.

Polyhedron. A solid bounded by plane polygons called the faces. Two faces interesect at an edge and three or more at a vertex. The five regular polyhedrons, the Platonic solids, in which all the faces are equal regular polygons, are the regular tetrahedron, hexahedron, octahedron, dodecahedron and icosahedron with 4, 6, 8, 12 and 20 sides respectively.

Polynomial. An expression of the form

$$a_0 x^n + a_1 x^{n-1} + a_2 x^{n-2} + ... + a_{n-1} x + a_n$$

where *n* is the degree of the polynomial. It is known as linear, quadratic, cubic, quartic, etc. When n is 1, 2, 3, 4, respectively.

Polytope. The analogue in n-dimensional space of point, segement, polygon, and polyhedron in spaces of dimensions 0, 1, 2 and 3. A **convex polytope** in n-space is the convex span of a finite set of point which do not all lie in the same hyperplane; thus a convex polytope is a bounded convex subset enclosed by a finite number of hyperplanes. A face of a convex polytope K is the empty set or K (the *improper faces*) or any set F for which there is a supporting hyperplane H of K with $F = H \cap K$. A facet of a convex polytope is a proper face not contained in any larger face.

Poncelet's Principle of Continuity. Refers to the very vague principle that "if one figure can be derived from another by a continuous change and the later is as general as the former, then any property of the first figure can be asserted for the second figure."

Pooled, Pooled Sum of Squares *(Statistics).* When several random samples of various sizes are regarded as originating from the same model, S is a **pooled sum of squares** if

$$S = \sum_{j=1}^{k} \sum_{i=1}^{nj} (x_i j - x_j)^2,$$

where j = i, ..., k over k samples and i = 1,...n_j where n_j is the number of observations in the jth sample and x_j is the mean of the jth sample . Then $S/\left[\sum_{j=1}^{k} nj\right]$ is a **pooled variance.**

Population. Refers to the whole set of items with some common characteristic. From the population, samples are taken for carrying out statistical analysis.

Position Vector. Refers to the vector than represents the displacement of a point from a given reference origin. If a point P in polar coordinates has coordinates (r, θ), then r is the position vector of P — a vector of magnitude r making an angle θ with the axis.

Positive. Means a number or quantity which is greater than zero. Numbers which are used in counting things and measuring sizes are all positive numbers. If a change in a quantity is positive, it increases, that is it moves away from zero if it is already positive and towards it if it is negative.

Positive Number. Refers to any number which is greater than zero. Such numbers are used, apart from other things, in counting things and measuring different sizes.

Poundage. (1) A weight expressed in pounds.

(2) A tax or charge calculated on the weight of chargeable goods in pounds or on their value in pounds sterling.

Pound Weight. The weight of one pound mass.

Power. (1) Rate of doing work. In F.P.S. system it is measured in foot pounds per second and in C.G.S. system in centimetre grams per second. In British unit. 1 horse power (hp) is 550 ft-lb/sec. (2) In algebra, it is the number of times a quantity is multiplied by itself. For example,

$$3^5 = 3 \times 3 \times 3 \times 3 \times 3,$$

here the power of 3 is 5 and in x^n the power of x is n.

Power Series. A series wherein the terms are regularly increasing powers of a variable. It is of the form

$$a_0 + a_1 x + a_2x^2 + a_3 x^3 + \ldots\ldots + a_n x^n + \ldots\ldots$$

where a_0, a_1, a_2 ... are constants.

Precession. If an object is spinning on an axis and a force is applied at right angles to this axis, then the axis of rotation can itself move around another axis at an angle to it. The effect is observed in tops and gyroscopes, which 'wobble' slowly while they spin as a result of the force of gravity. The Earth also precesses — the axis of rotation slowly describes a cone.

The precession of Mercury is a movement of the whole robit of the planet around an axis perpendicular to the orbital plane. It can be explained by relativistic mechanics.

Precision. The number of figures in a number. For example 2.342 is stated to a precision of four significant figures, or three decimal places. The precision of a number normally shows the accuracy of the value it represents.

Premium. Refers to the difference between the market value of the share and its face value. The share is said to be *above par, at par or below par* according as the market value of the share is more, equal, or less than its face value respectively.

Prime Number. An integer greater than one that has no integral factors except itself and one; for example 2, 3, 5, 7, 11, 13, 17. Every natural number can be expressed uniquely as a product of prime numbers; for example $1260 = 2^2 \times 3^2 \times 5 \times 7$. Prime numbers are of major interest in number theory.

Prime Polynomial. Refers to a polynomial expression that does not have any polynomial factor other than constants and itself. For example, $x + y$.

Prime as Symbol. Symbols $'$ and $''$ and $'''$ etc., placed at the top right of another symbol. Thus, (x', y') is a point different form (x, y) and (x'', y'') is a point different from (x', y'). The symbol is also used in derivatives. For example, a' represents first derivative and a'' represents second derivative of a.

Principal Diagonal. Elements of a determinant or a square matrix lying in a diagonal drawn from top left to bottom right. It is called as *Leading Diagonal.*

Principle of Undermined Coefficients. It states that if two polynomials in one varibale are equal, the coefficients of the terms of same power in the two polynomials must be equal. This principle is commonly used in finding the numerators in partial fractions.

Principle of Moments. It states that when a body is in equilibrium the sum of the moments of the forces in one direction must be equal to the sum of the moments of the forces in opposite direction.

Prism. A solid, generally made of a glass, with congruent parallel faces (called bases) and a set of rectangles (called lateral faces) obtained by joining corresponding vertices of the bases.

Probability. The mathematical concept concerned with the effects of chance on an event, experiment, or observation. It originated in 1654, when the mathematicians *Pascal and *Fermat worked on problems sent to them by a gambler.

If an event can occur in n ways and r is the number of ways it can occur in a specified way, then the **mathematical probability** of it occurring in the specified way is r/n. For example, the probability of the number 5 coming up on a six-faced dice in one throw is 1/6. The probability of a 5 coming up x times in x throws is obtained by multiplication, e.g. it is $(1/6)^x$. Probabilities are expressed as numbers between 1 (a certainty) and 0 (an impossiblity).

If in a number of experiments an event has occurred n times and failed m times, the **empirical probability** of success in the next trial is n/(n + m). Actuarial tables of life expectancy are based on empirical probabilities.

Program. Refers to a set of instructions to the computer, put in a sequential order, written in a programming language, special language understandable by the comoputer. The instructions are written in the binary number system. Instructions along with the relevant data on which the instructions operate are fed into the computer as input. The code that includes only binary numbers 0 and 1 is known as *binary code or machine code*. The programming language which is in machine code, is known as *machine language*. It is quite tedius to write program in machine code, hence the programs are written in *problem-oriented languages or high level language* like Fortran, Cobol, basic, etc., and are known as *source programs* which are translated by the complier into machine language . *Machine oriented languages or Assembly languages are all low level languages.*

Assembler changes assembly language programs into machine code.

Projectile. Refers to an object projected in space having a velocity at an angle to the horizontal line through the point of projection. The horizontal and vertical components of velocity are to be cosidered separately. The path trasversed by the projectile is always a parabola.

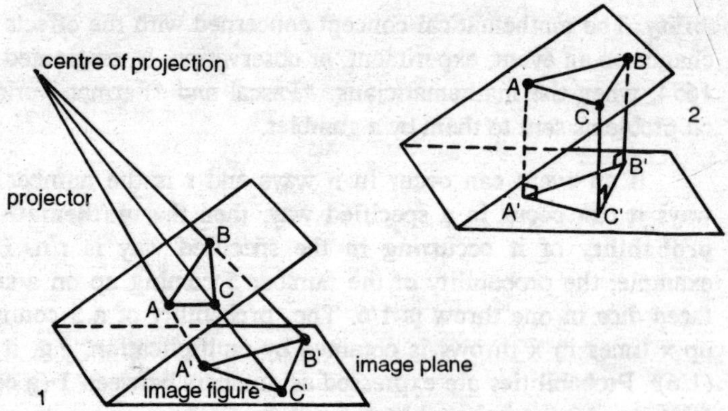

Fig. P-6

PROJECTION. *The mapping of a triangle by* (1) *central and* (2) *orthogonal projections.*

Projection (geometry). A mapping of a figure in one plane onto its image in another, called the image plane or plane of projection. A map of the earth's surface is an example of a projection. The image of each point in the figure is the intersection of a straight line (the projector) and the image plane. In central projection a projector is the straight line drawn through the point and a fixed point (the centre of projection). In orthogonal projection a projector is the perpendicular from the point to the image plane. The set of image points makes up the image figure.

Projective Geometry. Refers to the study of how to geometric properties of one configuration, are changed when subjected to projection by some geometric rules, to another configuration. *Central Projection* are those under which the properties of configuration subjected to such projection, do not get changed.

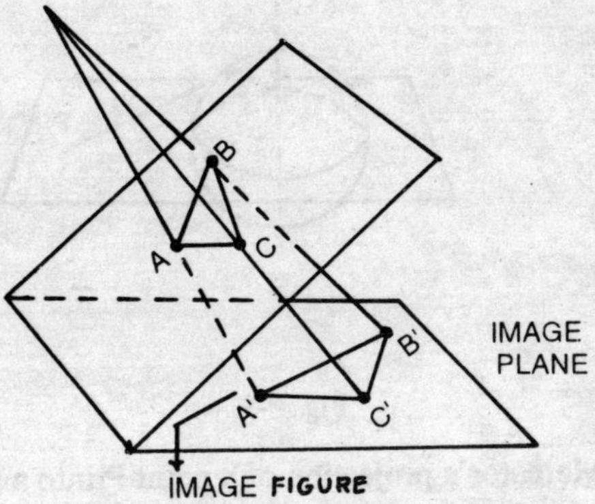

Fig. P-7.

Central projection of a triangle ABC onto a triangle A'B'C'.

Sterographic projection of a point P on the surface of a sphere onto a plane perpendicular to the line joining the poles N and S. The image P' is obtained by extending the line NP until it crosses the image plane.

Orthogonal projection of a triangle ABC onto a triangle A'B'C'

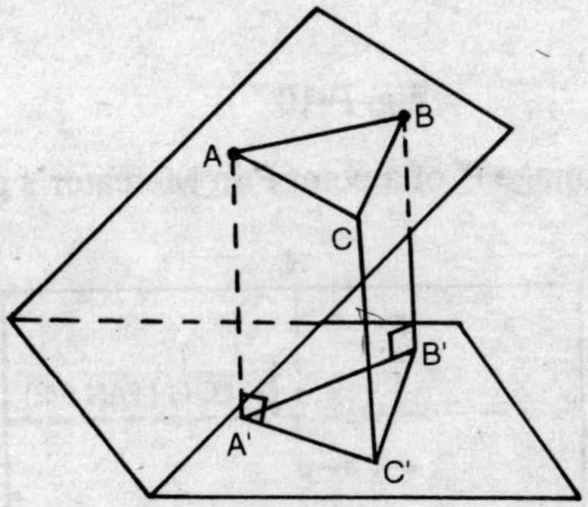

Fig. P-8.

Methods of projecting from one plane to another

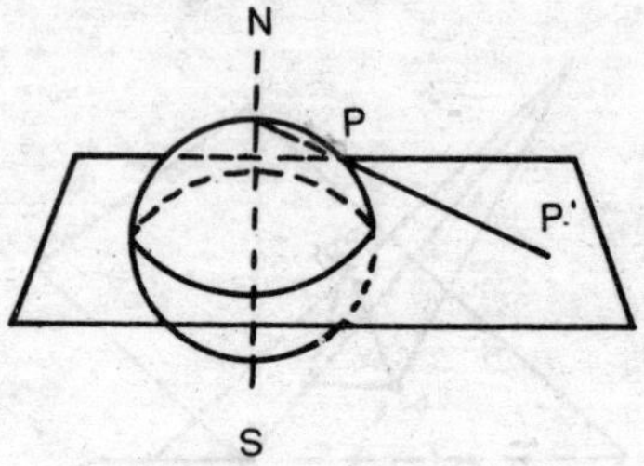

Fig. P-9.

Mercator's projection of a point P into a point P' in an image plane or map.

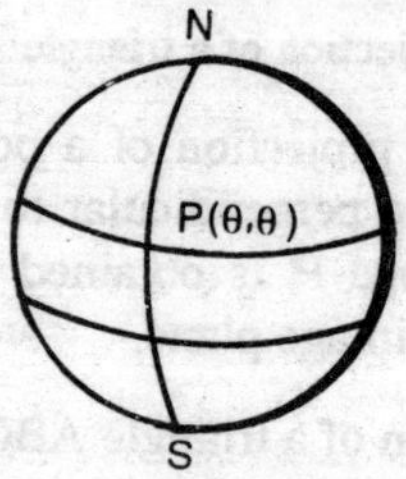

Fig. P-10.

The image P' of a point P in Mercator's projection.

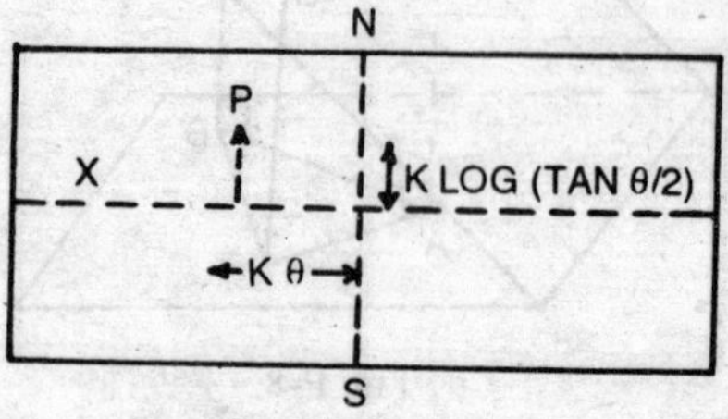

Fig. P-11.

Methods of projecting from the surface of a sphere onto a plane.

Proof. Means a logical argument indicating that a statement, proposition, or mathematical formula is true. A proof has a set of basic assumptions, known as axioms or premises, that are combined according to logical rules, to derive as a conclusion the formula that is being proved. A proof of a proposition or formula P is just a valid argument from true premises to P as a conclusion.

Proportion. Means a relation giving the equality of two ratio, in the form a/b = c/d. Earlier it was written as a:b: :c: d which has now become obsolete. The operation of *componendo, dividendo, and convertendo,* applied to the proportion a/b = c/d give,

$$\frac{a \pm b}{b} = \frac{c \pm d}{d}, \frac{a}{a \pm b} = \frac{c}{c \pm d}, \frac{a}{b} = \frac{c}{d} = \frac{a \pm c}{b \pm d}.$$

This is called as *simple proportion.*

Proportional. Symbol: α. Changing in a constant ratio to another quantity. For example, if the length l of a metal bar increases by 1 millimetre for evry 10° C rise in its temperature T, then the length is proportional to temperature and the constant of proportionality k is 1/10 millimetre per degree Celsius; $l = l_0 + kT$, where l_0 is the initial length. If two quantities a and b are *directly proportional* then a/b = k, where k, is a constant. If they are *inversely proportional* then their product is a constant: i.e., ab = k, or a = k/b.

Proposition. Refers to a sentence or formula in a logical argument. A proposition can have a truth value; that is, it can be either true or false but not both. Any logical argument consists of a succession of propositions linked by logical operations having a proposition as conclusion. Propositions may be simple or compound. A *compound preposition* is one that is made up of more than one preposition. For example, a proposition P might be made up of the constituent parts 'if R, then S or not Q' i.e., in this case P = R → (s ∨ Q). A *simple proposition* is one which is not compound.

Propositional Calculus. A system of formal logic that assigns truth or falsehood to compund propositions or statements on the basis of the rules governing the connective words between them. Examples of such connective words are "and," "or," or "if... then..." For example, the rule for "and" is that a statement containing it is true only if both the simple statements it connects are true; for "or" it is sufficient that only one statement be true. Thus "pigs fly and birds sing" is false, while "either pigs fly or birds sing" is true.

Ptolemaic System. A theory concerning the motions of the sun, moon and planets that was originally advanced in Greece in the 3rd century BC and was completed by Ptolemy in the 2nd century AD. It was a geocentric system in which the moon, Mercury, Venus, Sun, Mars, Jupiter, and Saturn moved around the earth. Prediction agreed reasonably well with observation. To achieve this Ptolemy had to propose that each orbiting body moved in a small circle (an epicycle) the centre of which moved in a larger circle (a deferent) around the earth. A sphere of stars lay beyond the seven deferents. The Ptolemaic system was finally abandoned when the *heliocentric system of Copernicus was accepted.

Ptolemy's Theorem. This establishes relations between the lengths of the sides and diagonals of a cyclic quadrilateral ABCD: AB.CD + BC.AD = AC.BD.

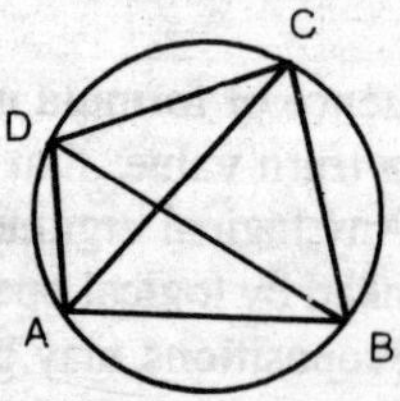

Fig. P-12.

Pure Mathematics. Whole of mathematics may be broadly classified into two categories: Pure Mathematics and Applied Mathematics. Pure Mathematics in that branch which deals with the mathematical theory and structures without intending to apply it to any field. Pure mathematics is abstract in nature.

Pyramid. It is a solid having base as a polygon and other faces are triangles having a common vertex. If the base is a regular polygon, a line from the vertex to the centre of the base is known as axis of the pyramid. The pyramid as a *right - pyramid* if its axis is perpendicular to its base, otherwise oblique. In a regular pyramid all the lateral faces are congruent isosceles triangles making the same angle at the base. A square pyramid has a square base and four congruent triangular faces. The volume of a pyramid is one-third the area of the base multiplied by its height.

Pythagoras (6th century BC). Greek philosopher and religious leader. Born at Samos, he migrated in about 530 BC to Crotone (S Italy), where he founded a religious society that governed Crotone for many years until its suppression (460-440 BC). Its members followed an ascetic regime of dietary taboos, self-examination, and study aimed at purifying the soul and releasing it from "entombment" in successive bodies. Pythagoras probably discovered the geometrical theorem named after him and certainly discovered the arithmetical ratios governing musical intervals, which led him to interpret the universe in terms of mathematics alone. Pythagoreans were pioneers in several branches of science.

Pythagoras Theorem. It states that in any right angled triangle the square on the hypotenuse is equal to the sum of the squares on the other sides.

Q

Quadrant. (1) Refers to one of four divisions of a plane. In rectangular Cartesian coordinates, the first quadrant is the area to the right of the y-axis and above the x-axis, where both x and y are positive. The second quadrant is the area to the left of the y-axis, where x is negative and y is positive. The third quadrant is below the x-axis and to the left to the y-axis, where both x and y are negative. The fourth quadrant is below the x-axis and to the right of the y-axis, where x is negative and y is positive. In polar coordinates, the first, second, third and fourth quadrants occur when the direction angle, θ, is 0 to 90° (0 to $\pi/2$); 90° to 180° ($\pi/2$ to π); 180° to 270° (π to $3\pi/2$); and 270° to 360° ($3\pi/2$ to 2π) respectively.

(2) A quarter of a circle, bounded by two perpendicular radii and a quarter of the circumference.

(3) A unit of plane angle equal to 90 degrees ($\pi/2$ radians). A quadrants is a right angle.

Quadratic Equation. Refers to a polynomial equation wherein the highest power of the unknown variable is two. The general form of a quadratic equation in the variable x is

$$ax^2 + bx + c = 0$$

where a, b and c are constants. It is also written in the reduced form

$$x^2 + bx/a + c/a = 0$$

In general, there are two values of x that can satisfy the equation. These solutions (or roots), are given by the formula

$$x = [-b \sqrt{} \pm (b^2 - 4ac)]/2a$$

The quantity $b^2 - 4ac$ is known as the discriminant. When it is a positive number, there are two real roots. When it is zero, there are two equal roots. When it is negative there are no real roots. The Cartesian coordinate graph of a quadratic function

$$y = ax^2 + bx + c$$

is a parabola and the points where it crosses the x-axis are solutions to

$$ax^2 + bx + c = 0$$

If it crosses the axis twice there are two real roots, if it touches the axis at a turning point the roots are equal, and if it does not cross it at all there are no real roots. In this last case, where the discriminant is negative, the roots are always two conjugate complex numbers

Quadrilateral. A plane figure having four straight sides. Its' six special case are given below:

(i) *Trapezium* (*or Trapezoid*). With two of its sides parallel.

(ii) *Parallelogram*. With both the pairs of its opposite sides parallel.

(iii) *Rectangle*. With all the vertical angles as right angles.

(iv) *Kite*. With two pairs of equal adjacent sides.

(v) *Rhombus*. With all the four sides equal. It can also be taken as a parallelogram having equal adjacent sides.

(vi) *Square*. With all the four sides equal and all the four vertical angles right angle. In can also be considered as a rectangle having equal adjacent sides.

Quart. A unit of capacity equal to two UK pints (in the UK) or two US pints (in the USA). In the USA a dry quart is equal to two US dry pints.

Quarter. Means one fourth. (1) A common fraction with 4 as denominator. (2) One of the phases of moon. (3) One fourth of a year, equal to three months.

Quartic Equation. A polynomial equation wherein the highest power of the unknown variable is four. The general form of a quartic equation in a variable x is

$$ax^4 + bx^3 + cx^2 + dx + e = 0$$

where a, b, c, d and e are constants. It is also written in the reduced form

$$x^4 + bx^3/a + cx^2/a + dx/a + e/a = 0$$

In general, there are four values of x that can satisfy a quartic equation. For example,

$$2x^4 - 9x^3 + 4x^2 + 21x - 18 = 0$$

can be factorized to

$$(2x + 3)(x - 1)(x - 2)(x - 3) = 0$$

and its soltuions (or roots) are — 3/2, 1, 2, and 3. On a Cartesian coordinate graph, the curve

$$y = 2x^4 - 9x^3 + 4x^2 + 21x - 18$$

crosses the x — axis at x = — 3/2; x = 1; x = 2; and x = 3.

Quartile. One of the three points which divide a set of data into four equal parts, wherein data has already been arranged in numerical order. The three quartiles are represented by Q_1, Q_2 and Q_3. Q_1 is 25th percentile P_{25}, Q_2 is 50th percentile P_{50} (median), and Q_3 is 75th percentile P_{75}.

Quintic Equation. Refers to polynomial equation wherein the highest power of the unknown variable is five. The general form of a quintic equation in a variable x is:

$$ax^5 + bx^4 + cx^3 + dx^2/a + ex + f = 0$$

where a, b, c, d, e and f are constants. It is also written in the reduced form

$$x^5 + bx^4/a + cx^3/a + ex/a + f/a = 0$$

In general, there are five values of x that satisfy a quintic equation. For example,

$$2x^5 - 17x^4 + 40x^3 + 5x^2 - 102x + 72 = 0$$

can be factorized to

$$(2x + 3)(x - 1)(x - 2)(x - 3)(x - 4) = 0$$

and its solutions (or roots) are $-3/2$, 1, 2, 3 and 4. On a Cartesian coordinate graph, the curve

$$y = 2x^5 - 17x^4 + 40x^3 + 5x^2 - 102x + 72.$$

Crosses x — axis at $x = -3/2$, $x = 1$, $x=2$, $x = 3$ and $x = 4$.

Quintillion. Means (1) In the U.S. and France, the number represented by one followed by 18 zeros or 10^{18}. (2) In England, the number represented by one followed by 30 zeros or 10^{30}.

Quotient, *n*. The quantity resulting from the division of one quantity by another. The division may have been actually performed or merely indicated; e.g., 2 is the quotient of 6 divided by 3, as is also 6/3. In case the division is not exact, one speaks of the *quotient* and the *remainder*, or simply the *quotient* (meaning the interger obtained plus the indicated division of the remainder), e.g., 7×2 gives the quotient 3 and the remainder 1, or the

quotient $3\frac{1}{2}$.

R

Radical Axis. Refers to the locus of a point from which two pairs of tangents of equal length can be drawn to two given circles. If $S_1 \equiv 0$, $S_2 \equiv 0$ represent the equations of two circles then $S_1 - S_2 = 0$ will give the equation of the radical axis. Thus for the circles

$$s_1 \equiv x^2 + y^2 + 2g_1x + 2f_1y + c_1 = 0$$

and
$$s_2 \equiv x^2 + y^2 + 2g_2x + 2f_2y + c_2 = 0$$

the radical axis is

$$2(g_1 - g_2)\,x + 2\,(f_1 - f_2)\,y + c_1 - c_2 = 0$$

If the two circles are intersecting the radical axis will be their common secant, and if they touch each other the common tangent is the radical axis.

Radical Centre. The radical centre of **three circles** is the point where the three radical axes of the circles, taken in pairs, intersect. This point is finite if, and only if, the centres of the circles do not lie on a line: The radical centre of **four spheres** is the point of intersection of the six radical planes obtained with respect to the six possible pairs of spheres formed from the given four. The point is finite if, and only if, the centres of the four spheres are not coplanar.

Radical Expression. Refers to an indicated root of a quantity. For example, $2\sqrt{3}$ or $\sqrt{3}$, $\sqrt[3]{y} (= y^{1/3})$, etc. $\sqrt{}$ is the *redical sign* and $\sqrt{}$ represents square root $\sqrt[3]{}$ represents cube root, $\sqrt[r]{}$ represents then n^{th} root or root with *index* n. In $\sqrt[r]{}$ x, x is called *redicand* or base.

Radical Plane to Two Spheres. The locus of the equation resulting from elimaintaing the square terms between the equations of the two spheres. When the spheres intersect, the radical plane is the plane of their circle of intersection.

Radius. Refers to the distance from the centre to any point on the circumference e.g., the distance from the centre of a sphere to any point on its surface is the radius of the sphere.

Radius of Gyration. If m_1, m_2, m_3,....... are the masses of particles of which a body of total mass M is made up of and r_1, r_2, r_3,.... are the distance of these particles from some axis then the moment of inertia I of the body about this axis is given by $I = MK^2 = m_1r_1^2 + m_2r_2^2 + m_3r_3^3 + ...$ The distance K is called radius of gyration of the body.

Ram. Random-access memory.

Random Access. Refers to a method of organizing information in a computer storage device such that one piece of information can be reached directly in about the same time as any other. Main store, disk units, and drum units all operate by random access and are therefore known as *random access-memory* (Ram). In contrast a magnetic tape unit operates more slowly by *serial access*: a particular piece of information can only be retrieved by working through the preceding blocks of data on the tape.

Random Sample (Statistical). Suppose results of an experiment are represented by a real number X. If the experiment is repeated in such a way that a particular outcome is not influenced by the other outcomes, then the set of outcomes $\{X_1, X_2,......X_n\}$ is a random sample of n. Tech. A **random sample of size n** of a random variable X is a set of n independent random variable $\{X_1, X_2,.....X_n\}$ each of these has the same distribution function as X. e.g., suppose an urn contains n balls numbered from 1 to n and that an event is the outcome of drawing a ball, the value of the associated random variable X being the number on the ball. If X_1 and X_2 are the numbers of the balls drawn on the first and second draw, respectively, then the set $\{X_1, X_2\}$ is a random sample of order 2 if the ball drawn on the first draw is replaced before the second draw.

Random Sequence (statistics). Refers to a sequence that is irregular, nonrepetitive, and haphazard. A **sequence of random digits** is a random sequence whose terms are chosen from the ten digits 0, 1,....,9 such that for each position the probability of a particular digit being chosen is 1/10 and the choices at two different places are independent. A random sample of k objects from n objects can be obtained by numbering the objects from 0 to n and then choosing a sequence of numbers from a sequence of random digits (each with the same number of digits an n), disregarding those greater than n, until k numbers have been chosen.

Random Variable. (Stochastic Variable.) Refers to a quantity that can accept any one of a number of the predicted values of the variable. A *discrete random variable* X has a definite set of possible values $x_1, x_2, x_3 \ldots x_n$ having probabilities $p_1, p_2 \ldots p_n$ attached to the values of the variable. Here $p_1 + p_2 + p_3 + \ldots + p_n = 1$. On the other hand, if X is a continuous random *variable* it can take any value in a given range. The probability of x having a particular values is known as the *probability density function* f(x) since if $x \in [a, b]$ then $\int_a^b f(x)\,dx = 1$, that is, the area under the curve represented by the probability density function f(x) is always 1.

Random Walk. A succession of " walks" along line segments wherein the direction (and possibly the length) of each walk is determined in a random way. Random walks can be used to get probabilistic solutions to mathematical or physical problems.

Tech. **A random walk** is a sequence $\{S_n\}$ where $S_n = \sum_{i=1}^{n} X_i$ and $\{X_n\}$ is a sequence of independent random variables. e.g., each X_i might have values 1 + h or - h having equal probability and the process could be seen as a person making a step of length h every r seconds with each step equally likely to be to the right or the left; then the probability at time t of being at distance x from where he was at time t = 0 can be given by a function U(x, t) that satisfies the difference equation

$$U(x, t + r) = 1/2\, U(x + h, t) + 1/2\, U(x - h, t).$$

The function U may be evaluated a approximately by letting a computing machine "make" a large number of random walks by reference to a sequence of random numbers. If $h^2 = r$ and $h \to 0$, the limit U (x, t) of U (x, t) satisfies the equation.

$$\frac{\partial^2 u}{\partial x^2} = 2\,\frac{\partial u}{\partial t}$$

with the boundary condition u(x, 0) = 0 if $x \neq 0$ and $\int_{-\infty}^{\infty} n\ (x,t)\ dx = 1$. A random-walk method is a type of Monte Carlo method.

Range. Means. (1) The difference between the largest and smallest values in a set of data. It is a measure of dispersion. In terms of percentiles, the range is $(P_{100} — P_0)$.

(2) A set of number of quantities which form possible results of a mapping. In algebra, the range of a function f(x) is the set of values that f(x) can take, for all possible values of x. For example, if f(x) is taking the square root of positive rational numbers, then the range would be the set of real numbers.

Rank. Refers to a method of ordering a set of objects in accordance with the magnitude or importance of a variable measured on them e.g., arranging ten men in order of height. If the objects are ranked using two different variables, the degree of association between the two rankings is given by the coefficient of rank correlation.

Rates. Refers to a tax paid to local government authorities by the owners of property. Each property is given a *rateable value*, that is assessed by a valuation officer. Each year the local authority declares its rate poundage, i.e., the amount the each rate payer should pay for each pound of rateable value of his property. Thus the amount of rates paid is determined each year by multiplying the rateable value by the rate poundage.

Rational Expression. An algebraic expression wherein variables do not occur with fractional indices. For example $3x^2 + 4x + 1$ is rational while $(3x^2 + 2)^{1/3}$ is not rational. It is an irrational expression.

Rationalized Units. Refers to a system of units wherein the equations have a logical form related to shape of the system. SI units forms a rationalized system of units. For example, in it formulae concerned with circular symmetry contain a factor of 2π; those concerned with radial symmetry contain a factor of 4π.

Rational Numbers. Symbol: Q. The set of numbers which includes integers and fractions. Rational numbers can be written down exactly as ratios or as finite or repeating decimals. For example, 1/3 (= 0.333...) and 1/4 (=0.25) are rational. The square root of 2 (=1.14142136...) is not .'

Reaction. According to Newton's third law of force whenever object A applies a force on object B. B applies the same force on A. An old word for force is 'action', 'reaction', is therefore the other member of the pair.

Often it is hard, or impossible, to decide what A and B are. Thus in the interaction between two electric charges, each exerts a force on the other. Thus, in general, action and reaction have little meaning. The word 'reaction' is still used in restricted cases, such as the reaction of a support on the object is supports. In this case the 'action' is the effect of the weight of the object on the support.'

Real Time. The actual time in which a physical process takes place or in which a physical process, machine, etc., is under the direct control of a computer. A *real-time system* is able to react sufficiently fast so that it may control a continuing process, making changes or modifications where ever necessary. Air traffic control and airline reservations require real-time systems.

Receipt, (Finance). Refers to (1) the act of receiving payment in money or goods. (2) A statement acknowledging money or goods that have been received.

Reciprocal. The reciprocal of a number is the number whose product with the given number is equal to 1; i.e., 1 divided by the number. For a fraction, the reciprocal is the fraction formed by intecharging the numerator and denominator in the given fraction. For any set of objects for which multiplication is defined and there is a multiplicative *identity* (whose product with any member x of the set is equal to x); the reciprocal (or *inverse*) of an object x is an object y such that xy and yx are each equal to the identity (provided there is only one y with this property), e.g., the reciprocal of the *polynomial* $x^2 + 1$ is $1/(x^2 + 1)$, since

$$(x^2+1), \quad \frac{1}{x^2+1} = 1 .$$

Reciprocal Curve of a Curve. Refers to the curve obtained by replacing each ordinate of a given curve by its reciprocal, the graph of the equation derived from the given equation (in Cartesian coordinates) by replacing y by 1/y. The graph of $y = 1/x$ and $y = x$ are reciprocals of each other; so are the graphs of $y = \sin x$ and $y = \operatorname{cosec} x$.

Reciprocal Equation. Refers to an equation in one variable whose set of roots remains unaltered if the roots are replaced by their reciprocals; an algebraic equation whose roots are unaltered if the variable is replaced, by its reciprocal, e.g., when x is replaced by 1/x and the equations are simplified, $x + 1 = 0$ yields $1 + x = 0$, and $x^4 - ax^3 + bx^2 - ax + 1 = 0$ yields $1 - ax + bx^2 - ax^3 + x^4 = 0$.

Reciprocal Polar Figures in the Plane. Two figures made up of lines and their points of intersection are reciprocal polar figures if each point in either one of them in the *pole* of a line in the other with respect to some given conic; **polar reciprocal triangles** are two triangles so that the vertices of each of them are the poles of the sides of the other with respect to some conic. **Polar reciprocal curves** are two curves so related that the *polar,* with respect to a given conic, of every point on one of them is tangent to the other.

Reciprocal System of Vectors. Sets of vectors A_1, A_2, A_3 and B_1, B_2, B_3 such that $A_i.B_i = 1$, $i = 1, 2, 3$ and $A_i.B_j = 0$ if $i \neq j$. If the triple scalar product

$$[A_1\ A_2\ A_3] \neq 0$$

then the set of vectors reciprocal to A_1, A_2, A_3 is $A_2 \times A_3 / [A_1\ A_2\ A_4]$, $A_3 \times A_1 / [A_1\ A_2\ A_3]$, $A_1 \times A_2 / [A_1\ A_2\ A_3]$.

Reciprocal Theorems. (1) *In plane geometry*, theorems such that the interchanging to two geometric elements, e.g., angles and sides, points and lines, etc., transfers each of the theorems into the other. Two such theorems are not always simultaneously true or false.

(2) *In projective geometry*, same as dual theorems.

Rectangle. A parallelogram having one angle a right angle and therefore all of its angles right angles; a quadrilateral having all right angles. A **diagonal** of a rectangle is a line joining opposite vertices; if the sides are of length a and b, the length of the diagonal is $\sqrt{a^2 + b^2}$. The **altitude** is the perpendicular distance from one side to the opposite side. The area of a rectangle is the product of two adjacent sides. If a rectangle has two sides of length 2 and 3, respectively, its area is 6.

Redeem. Means to repurchase, to release by making payments. To redeem a note, bond, or mortgage means to pay the amount it calls for. To redeem property means to get ownership by paying off lapsed liability for which it was security.

Redemption. Means the act of redeeming.

Reduced, Reduced Cubic Equation. Refers to a cubic equation of the form $y^3 + py + q = 0$; the form that the general cubic, $x^3 + ax^2 + bx + c = 0$, takes when the x^2 term is eliminated.

Reducible. A curve or surface is said to be **reducible** in a given region if it can be shrunk to a point by a continuous deformation without passing outside that region.

Reducible Polynomial. For some given domain or field, a polynomial that can be written as the product of two polynomials with degrees at least 1 and coefficients in the given domain or field.

Reduced Form (of a polynomial). The equation of the form

$$x^n + (b/a)\, x^{n-1} + (c/a)\, x^{n-2} + \ldots = 0$$

which is obtained from a polynomial of the form

$$ax^n + bx^{n-1} + cx^{n-2} + \ldots = 0$$

For example,

$$2x^2 - 9x + 12 = 0$$

is equivalent to the reduced form

$$x^2 - 9x/2 + 6 = 0$$

Both equations have the solution x = 3/2 and x = 4. The Cartesian coordinate graph of

$$y = 2x^2 - 9x + 12$$

crosses the x-axis at the same points, $x = 3/2$ and $x = 4$ as the graph of

$$y = x^2 - 9x/2 + 6$$

Reduction Formulae in Integration. Refers to formulae which reduce a given integral into a simple form. For example, the reduction formulae.

$$\int \sin^n x\, dx = -\frac{1}{n} \sin^{n-1} x \cos x + \frac{n-1}{n} \int \sin^{n-2} x\, dx$$

$$\int \cos^n x\, dx = -\frac{1}{n} \cos^{n-1} x \sin x + \frac{n-1}{n} \int \cos^{n-2} x\, dx$$

$$\int \sin^p \cos^q x\, dx = -\frac{1}{p+q} \sin^{p-1} x \cos^{q+1} x$$

$$+ \frac{p-1}{p+q} \int \sin^{p-2} x \cos^q x\, dx,$$

$$\int \sin^p \cos^q x\, dx = -\frac{1}{p+q} \sin^{p+1} x \cos^{q-1} x$$

$$+ \frac{q-1}{p+q} \int \sin^p x \cos^{q-2} x\, dx.$$

Reduction Formulae in Trigonometry. Refers to formulae or rather the identities which reduce a circular function of any angle in terms of acute angle. For example,

$\sin(90° \pm \theta) = \cos\theta$, $\sin(180° - \theta) = \sin\theta$,

$\sin(180° + \theta) - \sin\theta$, $\sin(360° - \theta) = -\sin\theta$,

$\cos(90° - \theta) = \sin\theta$, $\cos(90° + \theta) = -\sin\theta$,

$\cos(180° - \theta) = -\cos\theta$, $\cos(180° + \theta) = -\cos\theta$,

$\cos(360° - \theta) = \cos\theta$, $\tan(90° - \theta) = \cot\theta$,

$\tan(90° + \theta) = -\cot\theta$, $\tan(180° - \theta) = -\tan\theta$,

$\tan(180^0 + \theta) = \tan\theta$, $\tan(360^0 - \theta) = -\tan\theta$, etc.

Reflection. Refers to geometrical transformation of a point or a set of points from one side of a point, line, or plane to a symmetrical position on the other side. On *reflection in a line*, the image of a point P would be a point P' at the same distance from the line but on the other side. The line, the *axis of reflection*, is the perpendicular bisector of the line PP'. In a symmetrical plane figure there is an axis of reflection, also known as the axis of *symmetry*, where in the figure is reflected onto itself. An equilateral triangle, for example, has three axes of symmetry. In a circle, any diameter is an axis of symmetry. Likewise, a solid may undergo *reflection in a plane*. In a sphere, any plane passing through the sphere's centre would be a plane of symmetry.

In a Cartesian coordinate system, reflection in the x-axis changes the sign of the y-coordinate. A point (a, b) would become (a, — b). Reflection in the y-axis changes the sign of the x-coordinate, on reflection a point (a, b) would become (— a, b). In three dimensions, changing the sign of the z-coordinate is equivalent to reflection in the plane of the x and y-axis. *Reflection in a point* is equivalent to rotation through 180°. Each point P is moved to a position P' so that the point of reflection bisects the line PP'. *Reflection in the origin* of Cartesian coordinates changes the signs of all the coordinates. It is equivalent to reflection in the x-axis followed by reflection in the y-axis, or vice versa.

Region. Refers to a set that is the union of an open connected set and none, some, or all of its boundary points. The set is an **open region** if none of the boundary is included: it is a **closed region** if all of the boundary is included. For example, a **closed triangular region** (or **triangular region**) is a triangle along with its interior; a **closed circular region** (or **circular region**) is a circle along with its interior; a **closed rectangular region** (or **rectangular region**) is a rectangle along with its interior. The interior of a triangle, circle, or rectangle is an **open triangular region**, an **open circular region**, and an **open rectangular region**, respectively. The interior of a sphere is an open region, and the sphere along with its interior is a closed region.

Regression Line. A line y = ax + b, known as the regression line of y on x, that gives the expected value of a random variable y conditional on the given value of a random variable x. The regression line of x on y is not in general the same as that of y on x. If a scatter diagram of data points $(x_1, y_1),\ldots,(x_n, y_n)$ is drawn and a linear relationship is shown up, the line can be drawn by hand. The best line is drawn using the east squares method.

Relation. The relation R, from a set X to a set Y, is a proper subset of X. Y the cartesian product $\{(x, y)/x \in X, y \in Y\}$ such that xRy.

Relative. Expressed as a difference from or as a ratio to, some reference level. Relative density, for example, is the mass of a substance per unit volume expressed as a fraction of a standard density, like that of water.

Relative Density. *Syn. Specific Gravity.* The ratio of the density of a substance to that of water at 4^0 C in case of liquids and solids, and to that of hydrogen at 0^0 C at atmospheric pressure 760 mm in cases of gases.

Relative Error. Refers to the error or uncertainty in a measurement expressed as a fraction of the measurement. For example, if, in measuring a length of 10 metres, the tape measures only to the nearest centimetre, then the measurement might be written as 10 ± 0.01 metres. The relative error is $0.001/10 = 0.001$.

Relativistic Mass. Refers to the mass of an object as measured by an observer at rest in a frame of reference wherein the object is moving with a velocity v. It is given by

$$m_0 = m \sqrt{(1 - v^2/c^2)}$$

where m_0 is the rest mass, c is the velocity of light, and m is the relativistic mass. The equation is a result of the special theory of relativity, and is in agreement with experiment. No object can travel at the speed of light as its mass would then be infinite.

Relativistic Speed (relativistic velocity). Refers to any speed (velocity) that is sufficiently high so as to make the mass of an object significantly greater than its rest mass. It is generally expressed as a fraction of c, the speed of light in free space. At a speed of c/2 the relativistic mass of an object is about 15% greater than the rest mass.

Relativity. The theory or relativity was developed and proposed by Albert Einstein. In short we can say that all the natural processes are functions of time as well and time should be taken as fourth dimension. According to this theory there is nothing like *absolute value* of speed, it is always relative, *relative* to that of the observer.

Remainder Theorem. The theorem expessed by the equation

$$f(x) = (x - a)\, g(x) + f(a)$$

This means that is a polynomial in x, f(x), is divided by (x – a), where a is a constant, the remainder term is equal to the value of the polynomial when x = a. For example, if

$$2x^3 + 3x^2 - x - 4$$

is divided by (x — 4), then the remainder term is

$$f(4) = 128 + 48 - 4 - 4 = 168$$

The remainder theorem is very useful for factorization of a polynomial. In this example,

$$f(1) = 2 + 3 - 1 - 4 = 0$$

so (x — 1) is a factor.

Repeated Root. If f(x) is divisible by n $(x - a)^n$, than the equation f(x) = 0 will have x = a as a root repeated n times. For example, in the equation $x^5 + x^4 - 2x^2 + x + 1 = 0$, has — 1 as a triple root and + 1 as a double root.

Representative Fraction. A proper fraction showing the scale of a map or a plan. It is the ratio of the distance on the map to the distance on the ground, both expressed in the same units. For example, the representative fraction 1/1,000,000, represents the scale 1 cm ≅1,000,000 i.e., 1 cm on the map represents 10 km. on the ground, thus, in the representative fraction the numerator represents the distance on the map and the denominator the distance on the ground, both in the same units. Further, the representative fraction 1/63,360 indicates the scale 1 inch = 1 mile.

Reserve. (Life Insurance). Refers to the amount an insurance company requires (at a given time) to add to future net premiums and interest to pay all claims expected according to the particular mortality table being used (this is the difference between the present value of future benefits and the present value of future premiums and is a liability, also known as the **reinsurance fund** or **self-insurance fund**). The **reserve per policy** (the **net premium reserve**) is known as the **initial** reserve when computed at the beginning of a policy year just after the premium has been received, and the **terminal reserve** when computed at the end of a policy year before the premium has been paid. The average of the initial and teminal reserves is known as the **mean reserve**. The **prospective method** for computing reserves uses the fact that the reserve (initial or terminal) is the difference between the present value of future benefits and the present value of future premiums. It is also possible to use the past history of the policy to compute the reserve (the **retrospective method**.) computing the present value of the past difference of the *net level premiums* and *the natural premiums* (this difference is positive in the early years of the policy and negative in the later years). A **premium deficiency reserve** is the sum equal to the difference between the present value of future net premiums and future gross premiums (required in most states when gross premium is less than net premium).

Residue Class. In modulo (modular) arithmetic, all the positive integers can be partitioned into a number of residue classes, decided by the remainders after dividing the numbers by a given positive integer. For example, in module (3) arithmetic, one residue class consists of the numbers 1, 4, 7, 10, 13, 16,..... because on division each number leaves a remainder 1. A second residue class consists of 2, 5, 8, 11, 14,.... The number of such residue classes is infinite. Thus residue class is denoted by *residue class mod 3.*

Resolution of Vector. Means the process of determination of the components of a vector in two mutually perpendicular directions. These components are known as the resolved parts of the given vector. These component vectors when combined give the original vector, called resultant vector. Sometimes the term denote the process of resolving the given vector in the components in oblique directions.

Resonance. Refers to the large-amplitude vibration of an object or system when given impulses at its natural frequency. For instance, a pendulum swings with a natural frequency which depends on its length and mass. If it is given a periodic 'push' at this frequency—for example, at each maximum of a complete oscillation—the amplitude is increased with little effort. Much more effort would be required to produce as swing of the same amplitude at a different frequency.

Restitution Coefficient. It is the ratio of the relative velocities of two colliding bodies immediately after collision and their relative velocities before impact. If u_1 and u_2 are the velocities of the two bodies before collision and v_1, v_2 their respective velocitive after the impact, then the coefficient of restitution (e), is given by

$$e = \frac{v_1 - v_2}{u_1 - u_2}$$

In general $e < 1$.

A perfect elastic collision conserves the kinetic energy i.e., in such a case $e = 1$.

Resultant. A vector having the same effect as a number of vectors. So, the resultant of a set of forces is a force which has the same effect; it is equal in magnitude and opposite in direction to the equilibrant. Depending on the circumstances, the resultant of a set of vectors can be found by different method.

Retract. A subset X of a topological space T being a retract of T means that there is a continuous function f that maps T onto X and satisfies $f(x) = x$ if $x \in X$, i.e., the identity map of X has a continuous, extension to T. If X is a retract of T, then each continuous function on X has a continuous extension to T. A topological space X being an **absolute retract** means that if T is any normal topological space and X is homeomorphic to a closed subset Y of T, then Y is a retract of T. A *disk* or an n-*ball* is an absolute retract, but a *circle* is not.

Reuleaux "Triangle." Refers to a closed curve made up of three arcs which join vertices of an equilateral triangle; each are joins two vertices and lies on the circle through these vertices having its centre at the remaining vertex. This "triangle" is a *curve of constant width* in the sense that if r is the radius of the circles, then for any line L the "triangle' lies between two lines parallel to L and separated by distance r.

Reverse, Backward. Refers to a series of steps in a computation is taken in *reverse order* when the last it taken first, the next to last second, etc., a finite sequence of terms is in reverse order when the last is made first etc.

Revolve. To rotate about an axis or point. One would speak of revolving a figure in the plane about the origin, through an angle of a given size, or of *revolving* a curve in space about the x-axis with the understanding that the revolution is through an angle of 360° unless otherwise stipulated.

Rhombus. Refers to a plane figure having four straight sides of equal length. Its area is equal to half the product of the lengths of its two diagonals, which bisect each other perpendicularly. The rhombus is symmetrical about both of its diagonals and also has rotational symmetry, in that it can be superimposed on itself after rotation through 180° (π radians).

Riemann Sum. Refers to the series that approximates the area between the curve of a fuction f(x) and the x-axis:

$$\sum_{i=1}^{n} f(\zeta_i)\,\Delta x_i$$

where Δx is an increment of x, ζ, is any value of f(x) within that interval, and n is the number of intervals. The definite (or Riemann) integral is the limit of the sum as n becomes infinitely large and Δx infinitesimally small.

Right Angled Triangle. Refers to a triangle having one angle 90° and other two acute angles. The side, facing the angle 90° is called *hypotenuse*. If a and b are the sides of a right angled triangle and c the hypotenuse, then $c^2 = a^2 + b^2$.

Right Hand Rule. Consider two vectors $\vec{A}$ and $\vec{B}$. Stretch the right hand in such a way that the fingers point in the direction of the vector $\vec{A}$. rotate the hand in such way that the fingers now are set in the direction of the vector $\vec{B}$. Then the thumb indicates the direction of vector $\vec{C}$, where $\vec{C} = \vec{A} \times \vec{B} = |\vec{A}||\vec{B}| \sin \theta \hat{\eta}$.

Here θ is the angle between the vector $\vec{A}$ and $\vec{B}$ and $\hat{\eta}$ the unit vector in the direction of the product vector $\vec{C}$.

Right Handed Screw System. In the vector multiplication of two vectors $\vec{A} \times \vec{B}$, the vector $\vec{A}$ is considered of being moved towards the vector $\vec{B}$ and we think of a screw being moved in such a pattern. The direction of the ultimate motion of the screw determines the direction of the vector $\vec{A} \times \vec{B}$. The screw, when rotated anticlokwise, moves upward and when rotated clockwise moves downward. The principle explains the fact, $\vec{A} \times \vec{B} = -\vec{B} \times \vec{A}$.

Right Solid. Refers to a solid geometrical figure which is upright; for example, a cone cylinder, pyramid, or prism having an axis at right angles to the base.

Rigid Body. In mechanics, refers to a body for which any change of shape produced by forces in the body can be neglected in the calculations.

Rigid Motion. Means (1) Moving a configuration into another position, but making no change in its shape or size; a rotational transformation followed by a translation, or the two take in reverse order or simultaneously. An isometry is a rigid motion if it does not change the orientation of the coordinate (2). Superposition of figures in plane geometry is a rigid motion.

Rigidity, Modulus of Rigidity. Refers to the ratio of the shearing stress to the change in angle produced by the stress. *Syn.* shearing modulus.

Ring. A non-empty set R = (a, b, c,...) along with two binary operations + and × defined on its elements such that V a, b, c, ∈ R

1. The operation $^+$ is associative in R: $a^+(b^+c) = (a^+b)^+c$.

2. The operation $^+$ is Commutative in R: $a^+a = b^+a$.

3. There exists an identity element $0 \in R$ with respect to$^+$: $a^+ 0 = 0^+ a = a$.

4. These exists inverse — $a \in R$ for every element $a \in R$ such that $a^+(-a) = (-a)^+ a = 0$.

5. The operation $^+$ is associative in R: $a^+(b^+c) = (a^+b)^+c$.

6. The operation $^+$ is distributive with respect to the operation$^+$ in R: $a^+(b^+c) = (a^+b)^+(a^+c)$.

and $(b^+c)^+a = (b^+a)^+(c^+a)$.

The ring defined above is denoted by $(R, ^+, ^+)$.

Ring with Unity. A ring with an identity element (generally denoted by 1) with respect to the operation$^+$. A ring having unity is called a *Division Ring if* $\forall\, a \in R$ $(a \neq 0)$, there exists $a^{-1} \in R$, such that $a^+a^{-1} = a^{-1+}a = 1$.

Rolle's Theorem. If a function f(x) is continuous in a closed interval [a, b] and differentiable in the open interval] a, b [and f(a) = f(b) = 0, the there exists a points c ∈] a, b [such that f (c) = 0. Geometrically, if a continuous curve crosses the x-axis at two points x = a and x = b, and has a tangent at every point between x = a and x = b then there must be at least one point between a and b at which the tangent to the curve is parallel to the x-axis, i.e., the function f(x) has a maximum or minimum value between f(a) and f(b).

Root (algebra). (1) One of the equal factors of a number. The square root is one of two equal factors; for examle $\sqrt{9} = \pm 3$; the rth root of a number n is the number that when raised to rth power goves n. (2) the solution of an equation, i.e., the values of the variable that will satisfy the equation.

Root of Complex Number. If $z = r(\cos\theta + i\sin\theta)$, then $r^{1/n} = z^{1/n}$

$$\left[\cos\left(\frac{2k\pi + \theta}{n}\right) + i\sin\left(\frac{2k\pi + \theta}{n}\right)\right],$$ is known as nth root of z, where k = 0, 1, 2, ...n — 1 give the n roots of z.

Root of an Equation. The value (values) of x that satisfy the equation $f(x) = 0$ is (are) known as the root (roots) of the equation. In case of algebraic equation the number of roots is equal to the degree of equation.

Root of a Real Number. The solution of the equation $x^n = a$, is the nth root of the real number a. If $n = 2$, the root is known as the square root of a (denoted by $\sqrt{a}$), if $n = 3$, it is cube root of a and if $n = 4$, the fourth root and so on.

Rose. Refers to a curve whose polar equation is $r = a \text{ sing } n\,\theta$ or $r = a \cos n\,\theta$, where a is a real constant and n an integer. If n is odd the curve has n loops and when n is even the curve has 2n loops. The curves are called n — leaved or 2n — leaved rose.

Rotation. A rigid motion wherein a rigid body or a figure moves in such a way, that every point of the body or figure turns about a common centre in a plane or a common line in space. In case of rotation every point turns through the same angle. The rotation is positive if counterclockwise and negative if clockwise.

Rotational Motion. Refers to motion of a body turning about an axis. The physical quantities and laws used to describe linear motion all have rotational analogues; the equations of rotational motion are the analogues of the equations of motion (linear).

As well as the kinematic equations, the equations of rotational motion include $T = I\,\alpha$, the analogue of $F = ma$. Here T is the turning force, or torque (the analogue of force), I is the moment of inertia (analogous to mass), and α is the angular acceleration (analogous to linear acceleration).

The kinematics equations relate the angular velocity, ω_1, of the object at the start of timing to its angular velocity ω_2, at some later time, t, and thus to the angular displacement ϕ they are:

$$\omega_2 = \omega_1 + \alpha_t$$

$$\theta = (\omega_1 + \omega_2) - 2_t$$

$$\theta = \omega_1 t + at^2/2$$

$$\theta = \omega_2 t - \alpha\, t^2/2$$

$$\omega_2^{\,2} = \omega_2^{\,2} + 2\alpha\,\theta$$

Rotation of Axes. *Syn. Transformation of Axes.* In two dimensional coordinate geometry if the coordinate axes are rotated through an angle θ (keeping the origin same), then coordinates of a point P (x, y) with reference to the original axes change to (x′, y′) with reference to the new axes. The relationship between the transformed coordinates is given by

$$x = x' \cos\theta - y' \sin\theta \text{ and } y = x' \sin\theta - y' \cos\theta$$

Rounding (rounding off). Refers to the process of adjusting the least significant digit or digits in a number after a required number of digits has been truncated (dropped). This reduces the error due to truncation but still leaves a *rounding error* so that the accuracy of, say, the result of a calculation will be decreased. For example, the number 2.871 329 71 could be truncated to 2.871 32 but would be rounded to 2.871 33.

Routh's Rule. According to this rule the moment of inertia of a homogeneous body about any one of the three perpendicular axes of symmetry passing through the centre of mass is given by

(i) The product of the mass and one third of the sum of the squares of the other two semi-axes, in the case of a rectangular body. The moment of inertia of a uniform rectangular lamina of mass m, length l and breadth b about an axis perpendicular to its plane and passing through the centre of the mass is given by

$$m\left[\frac{(l/2)^2 + (b/2)^2}{3}\right] = m\left(\frac{l^2 + b^2}{12}\right)$$

(ii) The product of the mass and one fourth of the sum of the squares of the other two semi-axes, in the case of a circular or an elliptic body. The moment of inertia of a uniform circular disc of mass m and radius r about an axis perpendicular to its plane and passing through its centre is given by

$$m\left(\frac{r^2 + r^2}{4}\right) = \frac{mr^2}{2}$$

(iii) The product of the mass and one fifth of the sum of the squares of other two semi-axes in the case of a sphere or spheroid. The moment of inertia of a solid sphere of mass m and radius r about its diameter is given by

$$m\left(\frac{r^2+r^2}{4}\right)=\frac{2mr^2}{2}$$

Row Transformation. Refers to the following operations on a matrix.

(i) The intercharge of any two rows.

(ii) The multiplication of any row by a non-zero scalar.

(iii) The addition of a constant multiple of elements of any row to the corresponding elements of other row.

These operations if applied on columns of a matrix are known as elementary *column transformations*. The two transformations taken together are called elementary transformations.

Roulette. Means (i) The locus of a point that rolls without slipping along a fixed curve and is called *point roulette*.

(ii) The locus of the envelope of a straight line attached to a curve that rolls without slipping along a fixed curve and is known as a *line roulette*.

e. g., the cycloid, epicycloid etc.

Rule of Sarrus. A rule of expanding a determinant. Below the n^2 elements of nth order determinant, write the first (n - 1)

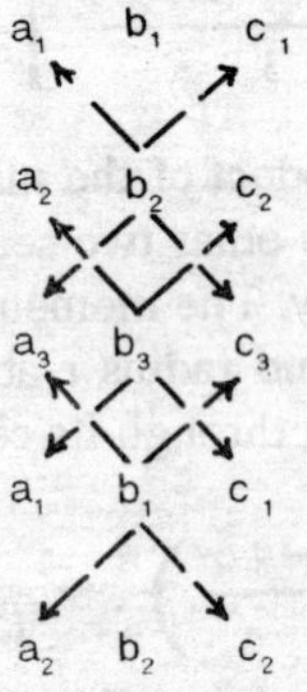

Fig. R-1.

rows as indicated in the figure. A descending line along the leading diagonal gives the first positive term, the remaining (n — 1) descending lines give other positive terms. A set of parallel ascending lines give negative terms. Thus, the value of the determinant

$$\begin{vmatrix} a_1 & b_1 & c_1 \\ a_2 & b_2 & c_2 \\ a_3 & b_3 & c_3 \end{vmatrix} = a_1b_2c_3 + a_2b_3c_1 + a_3b_1c_2 - a_3b_2c_1 - a_1b_3c_2 - a_2b_1c_3.$$

Runic Alphabet. An alphabetical writing system used for Norse and certain other Germanic languages. It originated, probably as a modified version of the Etruscan or the Roman alphabet, in the 2nd or 3rd century AD, and went out of use gradually after the 14th century. The Gothic word *runa* means secret or mystery, and runes were always associated with magical powers. The runic alphabet, called the *futhrark* from the names of the first six letters, consisted at first of 24 characters extended in Old English to 28 and later 31.

S

Saddle Point of a Matrix. Any finite matrix of real numbers, having elements a_{ij} in the ith row and jth column, may be considered to be the *payoff matrix* of a finite two—person zero—sum game. If the game has a saddle point at (i_0, j_0), then the matrix is said to have a saddle point at (i_0, j_0). A necessary and sufficient condition that a matrix have a saddle point is that there exist an element which is both the minimum element of its row and the maximum element of its column.

Sailing, Middle Latitude Sailing. Approximating the different in longitude (DL) of two place from their latitudes (L_1 and L_2) and departure (p) by the formula

$$p \sec \frac{1}{2}(L_1 + L_2) = DL$$

measured in minutes.

Parallel Sailing. Means sailing on a parallel of latitude; making use of the above formula, putting $L_1 = L_2$.

Plane Sailing. Means sailing on a rhumb line. The constant angle which the rhumb line makes with the meridians is known as the **ship's course**. Requires solving a plane right triangle.

Saint-Venant's Principle. If some distribution of force acting on a portion of the surface of a body is replaced by a different distribution of force acting on the same portion of the body, then the effects of the two different distributions on the parts of the body sufficiently far removed from the region of application of the forces are essentially the same, provided the two distributions of forces have the same resultant forces and moment.

Salient Point on a Curve. Refers to a point where two branches of a curve meet and stop and have different tangents. The curves

$$y = x/\left(1 + e^{1/x}\right) \text{ and } y = |x|$$

have salient points at the origin.

Sampling. Means the selection of a representative subset from a whole population. Analysis of the sample provides information about the whole population. This is known as *statistical inference.* For example, population parameters (like the population mean and variance) can be estimated by making use of sample statistics (like the sample mean and variance). Hypothesis tests are testing a new production process against an old. The population can be finite or infinite. In sampling with replacement each individual chosen is returned to the population before the next choice is made. In *random sampling* every member of the population has an equal chance of being chosen. In *stratified random sampling* the population is divided into strata and the random samples drawn from each are pooled. In *systematic sampling* the population is ordered, the first individual chosen at random and further individuals chosen at specified intervals, for example, every 100th person on the electoral roll. If a random sample of size n is the set of numerical values $\{x_1, x_2, \ldots x_n\}$, the sample mean is given by

$$\sum_{1}^{n} \bar{x} = x_i/n$$

The sample variance is given by

$$\Sigma (x_i - \bar{x})^2 / (n - 1)$$

$$\Sigma (X_i - \bar{x})^2/n$$

for a normal distribution. If μ is the population mean, the sample variance is given by

$$\Sigma (x_i - \mu)/n$$

Sampling Distribution. The distribution of a sample statistic. For example, when different samples of size n are taken from the same population the means of each sample form a sampling distribution. If the population is infinite (or very large) and sampling is with replacement, the mean of the sample means is $\mu_R = \mu$ and the standard deviation of the sample means is $\sigma R = \sigma/\sqrt{n}$ where μ and σ are the population mean and standard deviaton. When $n \geq 30$ sampling distribution are approximately normal and large sampling theory is used. When $n < 30$ exact sample theory is used.

Satisfy. To be a solution of, For example, 3 and — 3 satisfy the equation $x^2 = 9$.

Scalar Invariant. If $\vec{A}$ is a vector, then divergence $\vec{A}$, denoted by $\nabla . \vec{A}$, is known as scalar invariant.

Scalar Multiplication. The multiplication of a vector by a scalar quantity. Thus is $\vec{A}$ if a vector, taken $k\vec{A}$ represents a vector quantity whose magnitude is k times that of $\vec{A}$ and the direction is same as that of $\vec{A}$.

Scalar Product. Mean a multiplication of two vectors to give a scalar. The scalar product of A and B is defined by A.B. = AB cos θ, where A and B are the magnitudes of A and B and θ is the angle between the vectors. An example is a force F displaced s. Here the scalar product is energy transferred (or work done):

$$W = Fs$$

$$W = F.\ s \cos \theta$$

where θ is the angle between the line of action of the force and the displacement. A scalar product is shown by a dot between the vectors and is also known as a *dot product*. The scalar product is commutative

$$A.B = B.A$$

and is distributive with respect to vector addition

$$A.(B + C) = A.B. + A.C$$

If A is perpendicular to B, A.B = 0. In two-dimensional Cartesian coordinate with unit vectors i and j in the x- and y-directions respectively.

$$A.B = (a_1 i + a_2 j).\ (b_1 i + b_2 j) = a_1 b_1 + a_2 b_2.$$

Scalar Projection. Refers to the length of an orthogonal projection of one vector on another. For example, the scalar projection of A on B is A cos θ, or (A.B)/b where θ is the smaller angle between

A and B, and b is the unit vector in the direction of B.

Scalar Quantity. Means (1) The ratio between two quantities of the same kind, a number. (2) A number, as distinguished from a vector, matrix, quanternion, etc. (3) A tensor of order zero.

Scale Factor. The multiplying factor for each linear measurement of an object when it is to be enlarged about a given centre of enlargement. A scale factor can be positive, negative, or fractional. If the scale factor is positive, the image is larger than the object and on the same side of the centre of enlargement as the object. If the scale factor is fractional positive the image will be smaller than the object but on the same side of the centre of enlargement. If the scale factor is negative, the image will be on the opposite side of the centre of enlargement and will be inverted.

Scalene (1) Denoting a triangle having three unequal sides.

(2) Denoting a cone or cylinder having an axis not at right-angles to the base.

Scatter Diagram. Refers to a diagram that represent data from a bivariate frequency distribution (x_n, y_r). For example, the height and weight of n individuals represented as (x_1, y_2), (x_2, y_2).... (x_n, y_n), plotted on a graph paper, form the scatter diagram. A scatter diagram shows the dispersion of a set of data as well as a relationship between the two variables namely x and y. If the points on the graph are evenly distributed about a straight line they are said to be linearly corrected. If the points are distributed evenly around a curve, they are non-linearly correlated; otherwise uncorrelated.

Schwarz-Christoffel Transformation. Refers to a transformation that maps the interior of a polygon of the w-plane into the upper half of the z-plane (xy-plane) and the boundary of the polygon into the real axis. The transformation is

$$\frac{dw}{dz} = A\,(z - x_1)^{\frac{\alpha}{\pi} - 1}\,(z - x_2)^{\frac{\alpha}{\pi} - 1} \ldots\ldots (z - x_n)^{\frac{\alpha}{\pi} - 1}$$

where $\alpha_1, \alpha_2, ..., \alpha_n$ are the interior angles of the polygon with vertices $w_1, w_2, ..., w_n$ which map into the points $x_1, x_2, ..., x_n$ on the real axis of the z-plane.

Schwarz's-Inequality. Refers to the inequality

$$\left|\sum_{k=1}^{n} a_k b_k\right|^2 \leq \left(\sum_{k=1}^{n} |a_k|^2\right)\left(\sum_{k=1}^{n} |b_k|^2\right)$$

where, a_k, b_k (k = 1, 2, ..., n) are any complex numbers.

Scientific Notation (standard form). A number written as the product of a number between 1 and 10 with a power of 10. For example, 2342.6 in scientific notation is 2.3426×10^3, and 0.0042 is written as 4.2×10^3.

Screw. A type of machine that is related to the inclined plane, and, is practice, to the second order lever. The efficiency of screw systems is very low due to friction. Even so, the force ratio (F_2/F_1) can be very high.

The distance ratio is given by $2\pi r/p$, where r is the radius and p the pitch of the screw (the angle between the thread and a plane at right angles to the barrel of the screw).

Secant. Means (1) A line that intersects a curve. The intercept is a chord of the curve.

(2) (sec) A trigonometric function of an angle equal to the reciprocal of its cosine; i.e., $\sec \alpha = 1/\cos \alpha$.

Sector. Refers to (i) part of a circle that is bounded by two radii and the arc. If r is the radius of the circle and θ the angle in radians between the two radii, then the area of the sector is given by $\frac{1}{2} r^2 \theta$.

(ii) The solid generated by the revolution of a sector of a circle about a diameter of the circle.

Segment. Refers to (i) part of straight line or a curve between two of its points.

(ii) Part of a plane figure that is cut off by a straight line.

(iii) Part of a solid that is cut off by a plane.

Semi Circle. Refers to the part of a circle that is bound by a diameter and half the circumference of the circle. The angle in a semicircle is 90° or $\frac{\pi}{2}$ radians.

Semi Cubical Parabola. Refers to a curve that is represented by the equations $y^2 = ax^3$ or $x^2 = ay^3$.

Semi Latus Rectum. Means half the length of the latus rectum of a conic section.

Separation of Variables. Refers to a method used for solving differential equations. For example, if the differential equation $\frac{dy}{dx} = f(x, y)$ can be written as $\frac{dy}{dx} = \frac{f_1(x)}{f_2(y)}$, then the given equation can be solved as $\int f_2(y)\,dy = \int f_1(x)\,dx + c$.

Sequence. An order set of numbers, generally denoted by $a_1, a_2 \ldots a_r \ldots$, in which the rth term a_r can be expressed as a function of r.

Series. Refers to algebraic sum of the terms of a sequence. Each term of the series can be written as a function of its position in the series. A series $+ u_2 + u_3 + \ldots\ldots + u_n \ldots$ i.e, (Σu_n) is known as Convergent if the sum of its first n terms tends to a finite limit as $n \to \infty$ and divergent if the sum of its first n terms approaches infinity, as $n \to \infty$.

Set. Any collection of things or numbers which belong to, a well-defined category. For example, 'dog' is a member, or *element,* of the set of 'types of four-legged animal'. The set of 'days in the week' has seven elements. In set notation, this would be written as n {Monday, Tuesday,...} = 7, This type of set is a *finite set.* Some sets, like the set of natural numbers, N = {1, 2, 3, ...}, have an infinite number of elements. A line segment also is an infinite set of points.

Another way of writing a set of numbers is by defining it algebraically. The set of all numbers between 0 and 10 could be written as $\{x : 0 < x < 10\}$. That is, all values of a variable, x, such that x is greater than zero and less than ten.

Set Square. Refers to a drawing instrument, made up of a flat right-angled triangular shape, sued for drawing right angles and angles of 30°, 45° and 60°. Some types have a variable scale so that other angles can also be drawn.

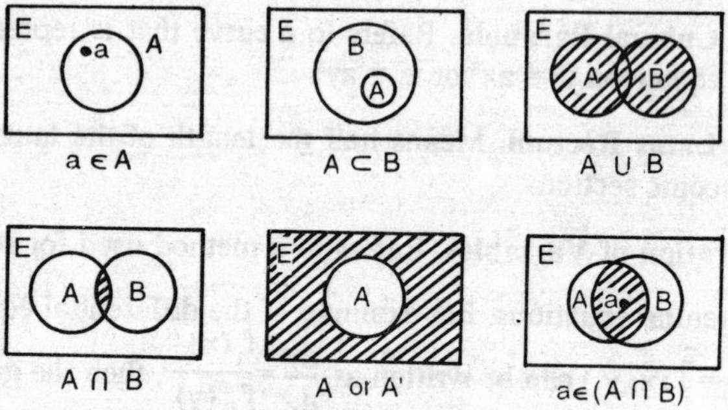

Fig. S-1.

SET THEORY: Venn diagrams.

Set Theory. The study founded by Georg Cantor, of the logical and mathematical laws of sets. A set is a defined collection of objects or elements; for example, the set of odd integers between 0 and 10 is {1, 3, 5, 7, 9}. The empty or null set, denoted by the symbol ϕ, has no elements. All sets are contained in the universal set E. The relationships between sets can be illustrated in a Venn diagram, named after the British logician John Venn (1834-1923), or shown by symbols. $a \in A$ means the element a is a member of the set A. $A \subset B$ means set A is contained in set B. $A \cup B$ means the union of A and B. $A \cap B$ means the intersection of A and B, i.e., those elements in both. A^c or A' is the complement of A, all elements in E but not in A.

Sexagesimal. Means based on multiples of 60. The measurement of an angle in degrees, minutes, and seconds, for example, is a sexagesimal measure, as there are 60 seconds in one minute, and 60 minutes in one degree. A sexagesimal number is one which uses 60 as a base instead of 10.

Significant Figures. Refers to the number of digits which are used to denote an exact value to a specified degree of accuracy. For exmaple, 6084.324 is a value accurate to seven significant figures. If it is written as approximately 6080, it is accurate to three significant figures. The final 0 is not significant because it is used only to show the order of magnitude of the number.

Similar. Means two or more figures that differ in scale but not in shape. Two triangles are similar if.

(1) Three sides of one proportional to three sides of the other.

(2) The angle of one is equal to the angle of the other and the sides forming the angle in one are proportional to the same sides in the other.

(3) Three angles of one are equal to three angles of the other.

Similar Figures. Refers to Geometrical figures which have same shapes but different size. More precisely, two or more figures are similar if a mapping can be defined between them such that the ratio of distances between pairs of points of one figure remains same for corresponding points of the figure.

Similar Teams. *Syn. Like Terms.* Refers to terms in an expression which differ in their coefficients only. Similar terms can always be combined into a single term. For example, $7x^2yz + 3x^2yz$ are like terms and hence combine to give a single term $10x^2yz$.

Similar Triangles. Two triangles are similar, if any one the following conditions are satisfied. (i) All the three angles of one triangles are equal to the corresponding angles of the other (ii) Three sides of one triangle are proportional to the corresponding sides of the other (iii) One angle of one triangle equal to the corresponding angle of the other and the side forming the angles of the first triangle are proportional to the corresponding sides of the other. The above three condition for similarity of triangles are known as (i) AAA postulate (ii) SSS postulates (iii) SAS postulate, respectively.

Simple Harmonic Motion (s.h.m.). Refers to any motion which can be drawn as a sine wave. For examples, the simple oscillation (vibration) of a pendulum or a sound source and the variation involved in a simple wave motion. Simple harmonic motion is observed when the system, moved away from the central position, experiences a restoring force that is proportional to the displacement from this position.

The equation of motion for such a system can be written in the form:

$$md^2x/dt^2 = -\lambda x$$

λ being a constant. During the motion there is an exchange of kinetic and potential energy, the sum of the two being constant (in the absence of damping). The period (T) is given

by $$T = 1/f$$
or $$T = 2\pi/\omega$$

where f is frequency and ω pulsatance. Other relationships are

$$x = x_0 \sin \omega t$$

$$dx/dt = \pm \omega \sqrt{(x_0^2 - x^2)}$$

$$d^2x/dt^2 = -\omega^2 x$$

Here x_0 is the maximum displacement; i.e., the amplitude of the vibration. In case of angular motion, like that for a pendulum, θ is used rather than x.

A simple harmonic motion can be shown by the motion of a point at constant speed in a circular path. The perpendicular from the point to an axis through a diameter describes a simple harmonic motion. This is used in a method of representing simple harmonic motions by rotating vectors known as *phasors.*

Simple Interest. Refers to the interest earned on capital, when the interest is withdrawn as it is paid, so that the capital remains fixed. If the amount of money invested (the principal) is denoted by P, the time in years by T, and the rate per annum in percent by R, then the simple interest is given by PRT/100.

Simpson's Rule. Refers to a rule for finding approximate area under a curve y = f(x). The area is divided into an even number of strips of equal width h, say. The curve is approximated by parabolic arc.

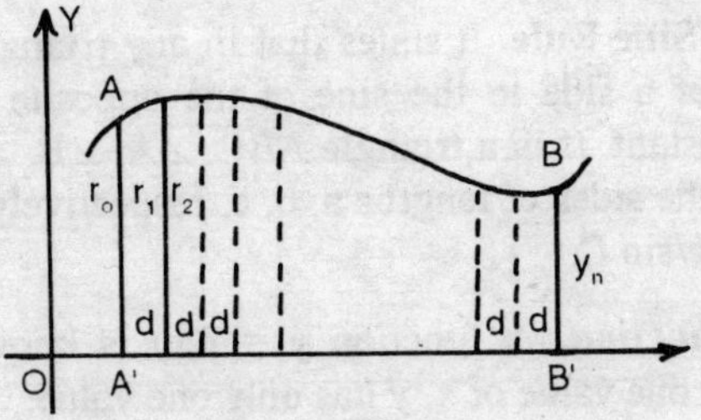

Fig. S-2.

The value of integral

$$\int_a^b y\,dx = \int_a^b f(x)\,dx$$

is the area ACOBA which is approximately equal to

$$\frac{h}{3}\left[\,y_0 + y_n + 4(y_1 + y_3 + \ldots.) + 2\,(y_2 + y_4 + \ldots..)\right]$$

This is considered to be a better approximation to the actual value than that by trapezoidal rule.

Simultaneous Equations. Refers to two or more equations which are true simultaneously and are thus satisfied by the same values of the unknowns involved. For example, the equations,

$$2x + y \;= 3$$

$$3x + 4y \;= 2$$

are simultaneous equations having solution $x = 2$, $y = -1$.

Sine Curve. Refers to the curve represented by the equation $y = \sin x$. It is a periodic curve with period 2π. Here $-1 \leq y \leq 1$.

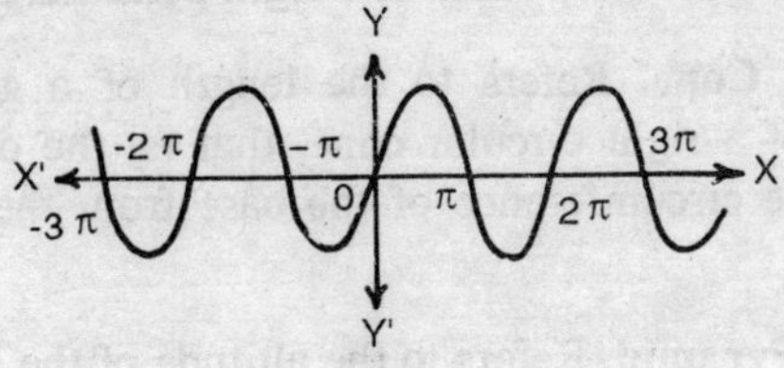

Fig. S-3.

The curve is also called a *sine wave*.

Sine Formula or Sine Rule. It states that in any triangle the ratio of the length of a side to the sine of the opposite angle always remains constant. If in a triangle ABC, ∠A, ∠B, ∠C are angles opposite to the sides of lengths a, b, c, respectively then a/sin A = b/sin B = c/sin C.

Single-Valued Function. A function y = f(x) is known as single-valued if for one value of x, y has only one value.

Singular Matrix. Refers to a square matrix whose corresponding determinant has the value zero. A singular matrix does not have inverse.

Singular Point. Refers to a point where a function of a complex variable ceases to be analytic.

Skew Lines. In space lines can be drawn which are neither parallel nor intersecting. The distance between any two of them is actually the shortest distance between them which is perpendicular to both the lines.

Skewness. Refers to a measure of the degree of asymmetry of a distribution. If the frequency curve has a long tail to the right (left) and a short one to the left (right), it is known as skewed to the right (left) or positively (negatively) skewed. Skewness is measured by either Pearson's first measure of skewness, (mean-mode divided by standard deviation, or the equivalent Pearson's second measure of skewness, divided by standard deviation).

Slant Height. Refers to (1) The length of an element of a right cone;

i.e., the distance from the vertex to the directrix.

(2) The altitude of the faces of a right pyramid.

Slant Height of Cone. Refers to the length of a generating line (element) of a right circular cone, that is, the distance of any point on the circumference of the base from the vertex of the cone.

Slant Height of Pyramid. Refers to the altitude of the lateral faces of a right pyramid.

Slide Rule. A mathematical instrument for performing multiplication and division. It consists of a rule that slides along a groove in another rule. Both rules are marked with logarithmic scales so that products and quotients can be calculated by effectively adding and subtracting *logarithms in the form of lengths of rule. Slide rules have now been largely replaced by pocket *calculators.

Small Circle. Refers to the section of the surface of a sphere by lane which does not pass through the centre of the sphere.

Smooth Surface. Refers to a surface for which the reaction between it and a body moving on it in any way is always along the normal to the surface, the friction between the surface and the body is to be neglected in the calculations.

Software. The suite of *programs that sets up a *computer system for operation as distinct from the physical equipment (hardware). Computer manufacturers generally supply machines complete with most of the basic software, which is regarded as an integral part of the computer. This controls various parts of the system and provides facilities for further programming. Supplementary programs to perform specific tasks are purchases separately or written by the user.

Solid Angle. Symbol: Ω. The three-dimensional analogue of angle. It is the region subtended at a point by a surface (rather than by a line). The unit is steradian (sr), which is defined analogously to the radian—the solid angle subtending unit area at unit distance. As the area of a sphere is $4\pi r^2$ the solid angle corresponding to the revolution (2π radians) is 4π steradians.

Solid Geometry. Refers to the study of geometric figures in three dimensions.

Solid of Revolution. Refers to a solid figure that can be formed by revolution of a line or curve (*the generator*) about a fixed axis. For example, rotating a circle about a diameter generates a sphere. Rotating a circle about an axis that does not cut the circle generates a torus.

Solution. Refers to a value of a variable that satisfies an algebraic equation. For example, the solution of $2x + 4 = 12$ is $x = 4$. An equation can have more than one solution; for example, $x^2 = 16$ has two; $x = -4$ and $x = +4$.

Solution of Triangles. Means calculating the unknown sides and angles in triangles. Because the sum of angles in a triangle is always 180°, the third angle can be calculated if two angle are known. All the sides and angles can be calculated when two sides and the angle between them is known, but if two sides and another angle is known there are two possible solutions. Any two angles and one side are sufficient to solve a triangle.

Space Curve. Refer to a curved line in a volume, defined in three-dimensional Cartesian coordinates by three functions:

$$x = f(t)$$

$$y = g(t)$$

$$z = h(t)$$

or by two equations of the form:

$$F(x, y, z) = 0$$

$$G(x, y, z) = 0.$$

Space-time Continuum. A coordinate system that has four dimensions, three representing physical space and the fourth time. The four-dimensional space-time continuum is used in *relativity to define an event. For example, an event occurring on the sun would be observed at different times on earth and on Jupiter, as light from the sun takes some 35 minutes longer to reach Jupiter than to reach the earth. Thus the concept of simultaneity requires a four-dimensional coordinate system to define events without ambiguity.

Spearman's Co-efficient of Rank Correlation. Refers to measurement of degree of association between the ranks of certain objects making use of two attributes x and y giving the data (x_1, y_1).... (x_n, y_n). The objects are ranked with respect to the two attributes first with respect to x and then with respect to y. di's denote the difference the ranks for each object. The Spearman's coefficient of rank correlation f is given by

$$f = 1 - \frac{6\,\Sigma d_i^2}{n\,(n^2 - 1)}.$$ See Rank.

Speed. Symbol: c. Refers to distance moved per unit time: $c = d/t$. Speed is a scalar quantity.

Usage can be confusing and it is common to meet the world 'velocity' where 'speed' is more correct. For example c_0 is the speed of light in free space, not its velocity.

Sphere. A closed surface made up of the locus of points in space that are at a fixed distance, the radius r, from a fixed point, the centre. A sphere is formed when a circle is turned through a complete revolution about an axis that is one of its diameters. The plane cross-sections of a sphere are all circles. The sphere is symmetrical about any plane that passes through its centre and the two mirror-image shapes on each side are known as hemispheres. In Cartesian coordinates, the equation of a sphere of radius r having its centre at the origin is given by

$$x^2 + y^2 + z^2 = r^2$$

The volume of a sphere is given by $4\pi r^3/3$ and its surface area is given by $4\pi r^2$.

Spherical Cap, Zone. The portion of the surface of a sphere lying between two parallel plane sections is known as *spherical zone.* A spherical zone wherein one plane is tangential to the spherical surface, is known as Spherical Cap.

Spherical Polar Coordinates. Refers to the polar coordinates which define the position of a point P in space and are given in terms of radial distance r from a fixed point O known as origin, the angle ϕ which is the angle made by the projection of OP on the XOY—plane with the axis of x, and the angle θ made by OP with the z-axis, that is $(r, \phi, \theta,)$. Suppose P is the cartesian coordinates (x, y, z) then

$$x = r \cos\theta \cos\phi,\ y = r \cos\theta \sin\phi,\ z = r \sin\theta.$$

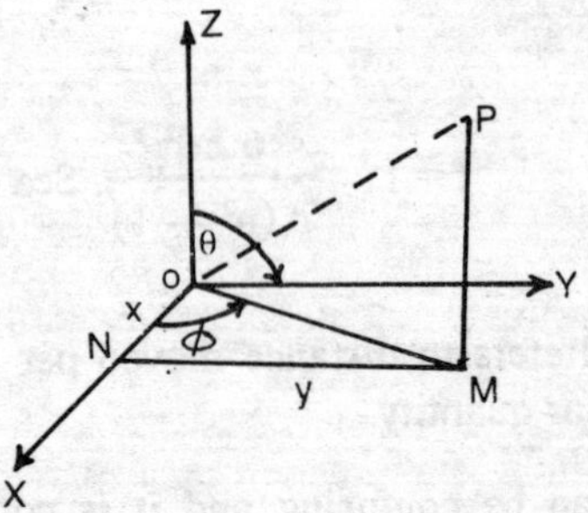

Fig. S-4.

Spherical Segment. The set of all the points equally distant from one fixed point known as centre. The common distance is known as radius of the surface. Actually, this spherical surface can be produced by revolving the arc of semi-circle about its bounding diameter. The area of the surface with radius r, is given by $4\pi r^2$ where r is the radius of the semi-circle.

Spherical Triangle. Refers to a portion of a spherical surface that is bound by the arc of three great circles. The sum of its spherical angles is 180 to 540.

Spherical Trigonometry. Refers to the study of properties of spherical triangle and their measurements.

Spiral. Refers to a curve described by the rotation of a point about a fixed point in such a way that its distance r, the radius vector, from the fixed point, the pole, is related to the angle of rotation θ, the vetorial angle. For example, the spiral of Archimedes $r = a\theta$, equiangular (logarithmic) spiral $r = ae^{\theta \cot \alpha}$.

Square. Refers to (1) The second power of a number of variable. The square of x is $x \times x = x^2$ (x squared). The square of the square root of a number is equal to that number. (2) In gemetry, a plane figure having four equal straight sides and right angles between the sides. Its area is given by the length of one of the sides squared. A square has four axes of symmetry i.e., the two diagonals, which are of equal length and bisect each other perpendicularly, and the two lines joining the mid-points of opposite sides. It can be superimposed on itself after rotation through 90°.

Square Root. Square root of a number may be defined as the number which when multiplied by itself produces the given number. It can be taken as the solution of the equation $x^2 - a = 0$ where a is the number given. If a is positive and real, x will have two real values having opposite signs and equal magnitude. If a is negative; x will have two imaginary values having equal magnitude but opposite signs.

Square Matrix. Refers to a matrix that has the same number of raws as that of columns, that is, a square array of numbers. The diagonal from the top left to the bottom right of a square matrix is known as the *leading diagonal* (or *principal diagonal*). The sum of the elements in the diagonal is known the *trace* (or *spur*) of the matrix.

Square Pyramid. *See* Pyramid.

$$\begin{pmatrix} 1 & 2 & 4 \\ 3 & 5 & 11 \\ 4 & 8 & 9 \end{pmatrix}$$

Fig. S-5.

A 3 × 3 square matrix. The sum of the elements in the leading diagonal will be 1 + 5 + 9 = 15.

Squaring the Circle. Refers to the attempt to construct a square which has the same area as a particular circle, using a ruler and compasses. An exact solution is impossible as there is no exact length for the edge, which is a multiple of the transcendental number $\sqrt{\pi}$.

Stability. A measure of how difficult it is to displace an object or system from its equilibrium.

Three cases are met in statics which differ in the effect on the centre of the mass of a small displacement. They are:

(1) *Stable equilibrium.* In this type of equilibrium the system returns to its original state when the displacing force is removed.

(2) *Unstable equilibrium.* In this type of equilibrium the system moves away from the original state when displaced a small distance.

(3) *Neutral equilibrium.* In this type of equilibrium when displaced a small distance the system is at equilibrium in its new position.

An object's stability can be improved by : (a) lowering the centre of mass; or (b) by increasing the area of support; or by both.

Stable Equilibrium. Refers to an equilibrium such that if the system is disturbed a little, there is a tendency for it to return to its original state.

Standard. Used as a reference.

(1) Equation written in standard form enables comparison with other equations of the same type. For example,

$$x^2/4^2 + y^2/2^2 = 1$$

and

$$x^2/3^2 + y^2/5^2 = 1$$

are equations of hyperbolas in rectangular Cartesian coordinates,

both written in standard form.

(2) A standard measuring instrument is one against which other instruments are calibrated.

Standard Deviation (σ). The square root of the mean of the squares of deviations from their average, used as a measure of dispersion in Statistics. It is equal to

$$\sigma = \frac{1}{n}\Sigma(x_i - \bar{x})^2$$

Standard form of Number. A number written as a number between 1 to 10. For instance, the number 2145.71 and 0.073534 can be written in standard form as 2.14571×10^3 and 7.3554×10^{-2}.

Standard Pressure. Refers to a pressure equal to mmHg at 0°C; 101325 Pa (approximately 100 kPa).

This is also known as one *atmosphere* (used as a unit of pressure). The *bar*, is used mainly when discussing the weather is 100 kPa exactly.

Statics. A branch of Mathematics that deals with the equilibrium of a body under the action of force. In the state of equilibrium the resultant force and torque are both zero.

Stationary Point. Refers to a point on a curve where the tangent will have zero slope and hence parallel to x-axis. All the turning points (maxima and minima) are stationary point.

At a stationary point the derivative obtained from the equation $y = f(x)$, will be zero. If the second derivative at the stationary point is negative then the maxima exists.

Stationary Wave. (standing wave). Refers to the interference effect due to two waves of the same type moving with the same frequency through the same region. The effect is often produced when a wave is reflected back along its own path. The resulting interference pattern is a stationary wave pattern. Here some points always show maximum amplitude; others show minimum amplitude. They are known as *antinodes* and nodes respectively. The distance between neighbouring node and antinode is equal to a quarter of a wavelength.

Statistics. The study of methods for collecting and analysing quantitative data. The data measure certain characteristics of a group of people or objects, called the population; usually the whole population cannot be observed, often because it is too large, so data are collected from a representative sample of the population. The sample is analysed and conclusions are inferred about the whole population, using *probability theory because the inferences cannot be certain. A population that has what is called a normal or Gaussian distribution (named after Karl *Gauss) varies randomly about a mean value. The standard deviation, the root-mean-square deviation from the mean value, is a measure of the spread in the population. The higher the standard deviation, the larger the sample size needed to be representative of the population. In binomial distributions the population consists of only two possible outcomes, as in tossing a coin, and in this case different statistical methods are used. In descriptive (or deductive) statistics data for a group are collected and analysed without conclusion being drawn about a larger group.

Steady-state Theory. A cosmological theory (see cosmology) proposed in 1948 by *Bondi, Hoyle and T. Gold (1920-) in which the universe is regarded as having always existed in a steady state. The *expansion of the universe is compensated by the continuous creation of new matter. On the present evidence this theory has largely been discredited in favour of the *big-bang theory.

Steradian. Symbol : sr. The SI unit of solid angle. The surface of a sphere, subtends a solid angle of 4π at its centre. The solid angle of a cone is equal to the area intercepted by the cone on the surface of a sphere of unit radius.

Stereographic Projection. Refers to a geometrical transformation of a sphere onto a plane. A point is taken on the surface of the sphere called the *pole* of the projection. The projection of points on the sphere onto a plane is obtained by taking straight lines from the pole through the points, and continuing them to the plane. The plane taken does not pass through the pole and is perpendicular to the diameter of the sphere through the pole.

Stirlings Theorem. According to it for very large values of n.

$$n! = n^n e^{-n} \sqrt{2\pi n}$$

approximately.

Stochastic Process. Refers to a process which produces a series of random values of a variable and hence a statistical distribution from them.

Stocks and Shares. Refers to the money borrowed by the government of a trading company is known as the stock. It is issued in the forms of bonds, promissory notes, securities, consols, debentures, etc. The shares issued by trading companies are another form of stocks. Persons holding these stocks are known as shareholders.

Profit divided among the share holder is known as dividend.

Store (memory). Refers to a system or device which is used in computing to hold information (programs and data) in such a way that any piece of information can automatically be retrieved by the computer as needed. The *main store* (or *internal store*) of a computer is under the direct control of the central processor. It is the area wherein programs, or parts of programs, are stored while they are being run on the computer. Data and program instructions can be extracted extremely rapidly by random access. The main store is supplemented by *backing store,* wherein information can be permanently store. The two basic forms of backing store are those in which magnetic tape is used (i.e., magnetic tape units) and those in which disks, or some other random-access device, are used.

The main store is divided into a large number of *locations,* each able to hold one unit of information, i.e., a word or a byte. The number of locations, i.e., the number of words or bytes that can be stored, gives the *capacity* of the store. Each location is identified by a serial number, called its *address*.

There are many ways in which memory can be classified. *Random-access momory* (RAM) and *serial access memory differ* in the way in which information is extracted from a store. With *volatile memory,* stored information is lost when the power supply is switched off, unlike *non-volatile memory.* With *read-only memory* (ROM), information is stored permanently or semipermanently; it cannot be changed by programed instructions but can in some types be changed by special techniques.

Stores may be magnetic or electronic in character. Magnetic tape units, disk units, and the recently developed bubble memory all are examples of magnetic memory. The electronic memory now widely used in main store is made up of highly complex integrated circuits. This *semiconductor memory* (or *solid state memory*) stores large quantity of information in a very small space; items of information can be extracted at very high speed.

Straight Line. The line that passes through two points in such a way that the line segment between the points is minimum. In general, the term line refers to straight line.

Stress. When external force is applied on an elastic body deformation takes place and due to that stresses develop. Stress can be conceived as internal forces per unit area in a body that has suffered strains. Stresses cannot be there in rigid bodies and can be witnessed in elastic, plastic bodies. Stresses may be *tensile* or *shear*.

Student's t-Distribution. Refers to the distribution, written t_n of a random variable

$$t = (\bar{x} - \mu)\sqrt{n}/\sigma$$

where a random sample of size n is taken from a normal population x with mean and μ standard deviation σ, n is known as the number of degrees of freedom. The mean of the distribution is 0 for $n>1$, and the variance is $n/(n - 2)$ for $n > 2$. When n is large t has an approximately standard normal distri bution. The probability density function, f(t), has a symmetrical graph. The value $t_n(\alpha)$ for which $P[t \leq t_n(\alpha)] = \alpha$ are tabulated for various values of n.

Student's t-Test. Refers to a hypothesis test for accepting or rejecting the hypothesis that the mean of a normal distribution having unknown variance is μ_0 making use of a small sample. The statistics $t = (\bar{x} - \mu_0)\sqrt{n}/s$ is computed from the data $(x_1, x_2 \ldots, x_n)$, where $\bar{x}$ is the sample mean, s is the sample standard division, and $n < 30$.

Subnormal-Subtangent. From any point P on a curve we draw tangent and normal which meet the x-axis at T and N in the figure,

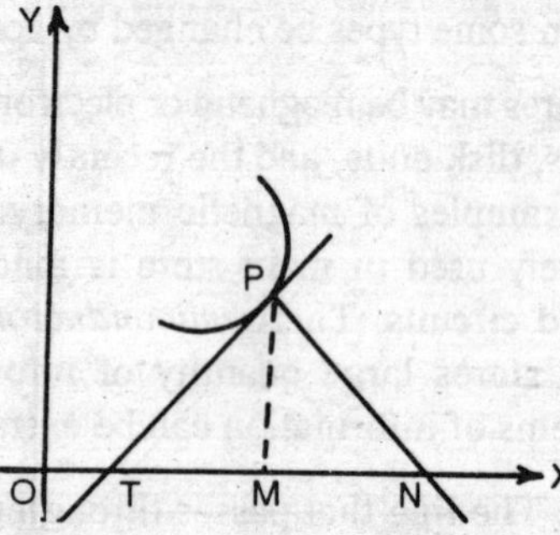

respectively. MN and TM are subnormal and subtangent for the curve at the point P. PT and PN are known as the lengths of tangent and normal at P. TM and MN can be considered as orthogonal projections of PT and PN on the axis of x respectively.

Subroutine. Refers to a section of a programme which requires repetitive use in different parts of the main programme to solve a problem by the computer. In the main programme, as soon as a subroutine is encountered the control is shifted to the subroutine with specified values of the input parameters. The subroutine has output paramters or arguments also wherein the results of computations during the execution of the subroutine are stored. Generally a program is made up of many subroutines which are extremely useful.

Subtend. Suppose the points A and B of a Chord AB of a curve have been joined to another point P on the curve. Then the arc AB or chord AB is said to subtend an angle $\angle APB$ at the point P. The diameter of a semi-circle always subtends a right angle at every point on the circumference. An arc or chord of a circle subtends an angle at the centre which is twice that at the circumference.

Subtraction (Symbol). Refers to a binary operation of finding the remainder when a number is removed from another. It is a reverse process of addition and can be described as finding the difference between two quantities. It is of the form $x - y$. It can be used to matrices, vectors, complex numbers as well.

Successive Differentiation. Means repeated differentiation to higher order derivatives. For example, if

$$\frac{dy}{dx} = f'(x), \text{ then } \frac{d^2y}{dx^2} = f''(x) \text{ etc.}$$

In general, nth order derivative represented as $\frac{d^ny}{dx^n}$ or $y^{(n)}$.

Summation Sign. Refers to the Greek capital letter Σ which is used to denote the summation of terms of a series. The sum of the first n terms of a sequence $\{u_r\}$ is given by

$$\sum_{r=1}^{n} u_r = u_1 + u_2 + ... + u_n.$$

whereas $\sum_{r=1}^{n} U_n$ stands for the sum of infinite number if terms of an infinite series.

Supplementary Angles. Refers to a pair of angles which add together to make a straight line (180^0 or π radius).

Supplementary Units. Refers to the dimensionless unit — the radian and the steradian — used along with base units to form derived units.

Surd. Refers to a numerical expression containing an irrational number. Surds are quadratic, cubic, quartic, etc., depending upon the power involved, i.e., 1/2, 1/3, 1/4 etc. For example, $\sqrt{3}$, $\sqrt{7}$.

These are also called *pure surds* while a *mixed surd* involves a rational number as well. For example $2 + \sqrt{3}$ or $5\sqrt{\pi}$

Surface. Refers to a set of points which form a space with two dimensions only. Surface may be plane like that of circular plate or curved like that of a sphere.

Surface of Revolution. Refers to a surface generated by rotating a curve about an axis. For example, rotating a parabola about its axis of symmetry forms a paraboloid of revolution.

Surveying. The measurement of distances, levels, angles, etc., on above, or below the earth's surface. It is necessary for the delineation of property lines (**boundary surveying**) and for planning the construction of almost any building or structure. **Plane surveying** is suitable for smaller areas as it neglects the earth's curvature, whereas **geodetic surveying** does not. Topographic surveys locate natural features and elevations for map making, dam planning, etc.

Symbolic Logic (formal logic). Refers to the branch of logic wherein arguments, the terms used in them, the relationships between them, and the various operations that can be performed on them are all represented by symbols. The logical properties and implications of arguments can then be more easily studies strictly and formally, making used of algebraic techniques, proofs, and theorems in a mathematically rigorous way. It is also known as mathematical logic.

The simplest system of symbolic logic is *propositional logic* (also known as *propositional calculus*) wherein letters, e.g., P, Q, R, etc., stand for proposition or statements, and various special symbols stand for relationships that can hold between them.

Symmetrical. Representing any figure that can be divided into two parts which are mirror images of each other. The letter A, for example, is symmetrical, and does not change when viewed in a mirror, but the letter R is not. A symmetrical plane figure has at least one line that is an axis of symmetry, which divides it into two mirror images.

Symmetric Function. A function in two or more variables which remains unchanged when any two variables are interchanged. For example, the function $x^3 + y^3 + z^3 - 3x^2y^2z^2$.

Symmetric Function of Roots. Suppose $\alpha_1, \alpha_2, \ldots\ldots, \alpha_n$ be the roots of the equation $a_0x^n + a_1x^{n-1} + a_2x^{n-2} + \ldots\ldots x_n = 0$. Then the sum of the roots taken one at a time will be $-a_3 / a_0$, taken two at a time is a_2/a_0; taken three at a time will be $-a_1/a_0$, and so on. Thus, for a quadratic; $\alpha_2 + \alpha_2 = -a_1 / a_0$, and for a cubic $\alpha_1 + \alpha_2 + \alpha_3 = a_1/a_0$, $\alpha_1\alpha_2 + \alpha_2\alpha_3 + \alpha_3\alpha_1 = a_2/a_0$, $\alpha_1 \alpha_2 \alpha_3 = -a_3/a_0$.

These are all called symmetric functions of the roots.

Symmetric Relation. A relation R defined on a set S is known as symmetric if for any two elements a and b of S, a R b implies bRa.

Symmetry. Two points A and B are said to be symmetrical or are said to have symmetry about a plane, a line or a point, if A is reflection of B and also, B is a reflection of A about the plane, the line or the point.

System Analysis. Refers to the thorough study of the working of an organisation or a system, the purpose for which it is meant and its basic requirements in order that a system analyst may, improve upon the performance of the system. The system analyst is expected to reduce the total number of steps involved.

T

T, *t*-Distribution (Statistics). A random variable X has a **t-distribution** or is a **t -random variable** having n degress of freedom if it has *probability density function f* for which

$$f(x) = \frac{\Gamma[\frac{1}{2}(n+1)]}{\sqrt{n\pi}\,\Gamma(\frac{1}{2}n)}\left(1+\frac{x^2}{n}\right)^{-1/2(n+1)}$$

where F is called the *gamma function*. The *mean* is 0 if n > 1 and the *variance* is n (n–2) if n > 2. A random variable is a random variable if it is symmetrically distributed about 0 and its square is an F *random variable* with (1, n) degrees of freedom. Equivalently, $(X-\mu)\sqrt{n/s}$ is a t-random variable if X is a normal random variable with mean μ and

$$s = \left[\sum_{i=1}^{n}(X_1-\mu)^2\right]^{1/2}$$

where $(X_1.......X_n)$ is a random sample of X. For large values of n, the t-distribution is approximately the normal distribution having mean 0 and variance 1. The t-test and t *distribution* were given by 'Student' (W.S. Gosset), although the distribution had been noted earlier by Helmert. *Syn*. Student's t.

t-Test. Suppose X is a normal random variable having unknown mean and variance. To test (with level of significance α) the hypothesis that the mean is μ_0 the t test uses the random variable

$$T = \frac{(n-1)^{1/2}(\mu-\mu_0)}{s},$$

where μ is the mean of a random sample of size n and s is $\left[\sum_{i=1}^{n}(X_i-\mu)^2/n\right]^{1/2}$; when the hypothesis is true, T has the t distribution with n – 1 degree of freedom and the hypothesis is rejected if $| T | > t_\alpha$, where $F(t_\alpha) = 1 - ½\alpha$ with F the distribution function for this t distribution. The t distribution can be used for many other tests of hypothesis.

Table. Refers to a systematic listing of results already worked out, which reduces the labour of computors and investigators and forms a basis for future predictions.

Tabular, Tabular Differences. Refers to the differences between successive values of a function, as recorded in a table. The tabular differences of a table or *logarithms* are the differences between successive mantissas, generally recorded in a column of their own. The tabular differences of a *trigonometric tables* are the differences between successive recorded values of trigonometric function.

Tac-Point. A *tac-point* of a family of curves is a point at which two different members of the family intersect and have a common tangent; a *tac-locus* is a set of tac-points, e.g., for the family of circles of radius one which are tangent to the x-axis, the lines $y = 1$ and $y = -1$ are tac-loci.

Tanget. The straight line that touches a curve at only one point, known as the point of contact. A tangent plane is one that touches a curved surface at one point.

Tangent Line. It is a line at a given point P on a curve is the limiting position of a secant PQ when the point Q approaches the point P along the curve.

Tangent Plane. A plane touching a surface at a point P known as the point of contact, so that any line through P in this plane is a tangent line to the surface at the point P.

Tension. Refers to internal force of a body which results in its stretch, that is resisting the external forces applied to the body to increase its length.

Tensor. A generalization of a vector of two or three dimensions to n-dimensional entity. Tensors depict how all the components of a vector quantity behave while undergoing certain transformations.

Term. (1) A term of a **fraction** means the numerator or denominator of the fraction. A term of a **proportion** means any one of the *extremes or means.* (2) A term of an **equation** or an **inequality** is the entire quantity on one (or the other) side of the sign of equality or inequality. *Member* is better usage. (3) For an expression that is written as the sum of several quantities, each of these quantities is known as a term of the expression, e.g., in

$xy^2 + y \sin x - \dfrac{x+1}{y-1} - (x + y)$, the terms are xy^2, $y \sin x$, $-\dfrac{x+1}{y-1}$ and $-(x + y)$. In **polynominal** $x^2 - 5x - 2$, the terms are x^2, $-5x$, and -2; in $x^2 + (x + 2) - 5$, the terms are x^2, $(x + 2)$, and -5. A **constant** (or **absolute**) term is a term which does not contain any of the variables. An **algebraic term** is a term that contains only algebraic symbols and numbers; e.g., $7x$, $x^2 + 3ay$, and $\sqrt{3x^2 + y}$. An algebraic term is **rational and integral** if the variable or variables do not appear under any radical sign, in any denominator, or with fractional or negative exponents. The terms $2xy$ and $2x^2y^2$ are rational and integral in x and y. A term that is not algebraic is **transcendental**. For examples transcendental terms are **trigonometric terms,** which contain only trigonometric functions and constants (sometimes a term is known as trigonometric when it also contains algebraic factors); **exponential terms,** are those terms that contains the variable only in exponents (sometimes terms are known as exponential when they contain both exponential and algebraic factors); **logarithmic terms** are those which contain the variable or variables affected by logarithms, such as $\log x$, $\log (x + 1)$ (sometimes terms are known as logarithmic when they also contain algebraic factors, as $x^2 \log x$). **Like** (or **similar**) terms are terms which contain the same variables, each variable of the same kind being raised to the same power; $2x^2yz$ and $5x^2yz$ are **like terms**. (4) In **finance** term means a period of time.

General Term. Refers to a term containing parameters in such a way that the parameters can be given specific values so that the term itself reduces to any special term of some set under consideration. The general term in the general algebraic equation of the nth degree $a_0x^n + a_1x^{n-1} + ... + a_n = 0$ would be written a_ix^{n-1} $(i = 0,..., n)$.

Terminal. Refers to a device which a computer user can have a direct access to the computer even while sitting at a distance. By terminal the user feeds the input in the form of data and instruction and gets the output also in the form of results. The input is fed in the computer through a keyboard similar to that of a typewriter, displayed on the screen, output also appears on the screen. One can make changes in the instructions and data and correct the syntax errors etc.

Terminating. Coming to an end; limited; expressed in a finite number of figures or terms e.g., the decimal 3.147 is a **terminating decimal,** while the repeating decimal 7.414141.... is **non-terminating.** Infinite sequences and series are non-terminating.

Ternary, Ternary Number System. Refers to a system of notation for real numbers that makes use of the base 3 instead of 10, e.g., the number $38\frac{5}{27}$ in the decimal system would be 1102.012 when written with base 3, since with base 3 the number 1102.012 is equal to $1.27 + 1.9 + 0.3 + 2 + 0.\frac{1}{3} + 1,\frac{1}{9} + 2,\frac{1}{27}$.

Ternary Operation. Refers to an operation which is applied to three objects. Examples are the operation of determining the *median* of three numbers, and the operation on the numbers x, y and z that produces the number x(y + z). *Tech.* A **ternary operation** is a function f whose domain is a set of ordered triples of members of a set S.

Terrestrial, Terrestrial Triangle. A spherical triangle on the earth's surface (considered as a sphere) having for its vertices the north pole and two points whose distance apart is being found.

Tetrahedral Surface. A surface admitting parametric representation of the form

$$x = A\,(u-a)^{\alpha}\,(v-a)^{\beta},$$

$$y = B\,(u-b)^{\alpha}\,(v-b)^{\beta},$$

$$z = C\,(u-c)^{\alpha}\,(v-c)^{\beta},$$

where a, b, c, A, B, C, α and β are constants.

Theorem. Refers to a general conclusion based on certain assumptions which are to be proved or a logical conclusion, proof of which is already known.

Theory. Refers to the principles concerned with a certain concept, and the facts postulated and proved about it.

Function **Theory**. Refers to the theory of functions of a real variable, the theory of functions of a complex variable, etc.

Group **Theory, or Theory of Groups**. See group.

Number **Theory, or Theory of Numbers** (Elementary). Deals with the study of integers and relations between them.

Theory of Equations. Deals with the study of methods of solving and the possibility of solving polynomial equations, and of the relations between roots and coefficients of equations.

Theory of Games. Refers to theory of probabilities of varying outcomes in games. These are calculated from the rules of the game and the number of players, making use of techniques of statistics.

Theory of Indices. It provide the laws for multiplication and division of quantities which are powers of the same quantity, expressed as under

$$a^p a^q = a^{p+q},\ a^p/a^q = a^{v-q},\ (a^p)^q = a^{pq}.$$

$$n \sqrt{a^m} = a^{m/n},\ a^1 = a,\ a^0 = 1,\ a^{-m} = 1/a^m.$$

Theta Functions. Let $q = e^{\pi/\tau}$, where t is a constant complex number whose imaginary part is positive. The four theta functions (generally written without explicitly indicating the dependence on τ) are

$$\theta_1(z) = 2\sum_0^{\infty}(-1)^n q^{(n+1/2)2} \sin(2n+1)z,$$

$$\theta_2(z) = 2\sum_0^{\infty} q^{(n+1/2)2} \cos(2n+1)z,$$

$$\theta_3(z) = 1 + 2\sum_1^{\infty} q^{n^2} \cos 2nz,$$

$$\theta_4(z) = 1 + 2\sum_1^{\infty} (-1)^n q^{n^2} \cos 2nz.$$

Other notations are also used for these functions (e.g., θ for ∂_4) and $\theta_i(\pi z)$ for $\theta_i(z)$. It can be shown that $\theta_1(z) = -\theta_2(z + \frac{1}{2}\pi) = -iq^{1/4} e^{iz}) \theta_3(z + \frac{1}{2}\pi + \frac{1}{2}\pi\tau) = (-iq^{1/4} e^{iz}) \partial_4(z + \pi\tau)$. Each of the theta functions satisfies a relation analogous to

$$\theta_4(z+\pi) = \theta_4(z) = (-qe^{2iz})\,\theta_4(z+\pi\tau),$$

and are said to be *quasi doubly-periodic*. The theta functions are entire functions.

Three, Rule of Three. Refers to the rule that the product of the means of a proportion equals the product of the *extremes* (see proportion). This rule enables one to find any one of the numbers of a proportional if the other three are given. If, $3/x = 2/5$ then $2x = 15$ and $x = 7\frac{1}{2}$.

Three Dimensional Geometry. Deals with the study of figures in three (as well as two) dimensions.

Three-Point Problem. Given three collinear points, A, B, C, with the distances AB and BC known, and a fourth point S, with the angles ASB and BSC known, to find the distance SB. This is the problem of finding the distance from a ship S to a point B on the shore.

Tietze Extension Theorem. If T is a Hausdorff topological space, each of the following is a necessary and sufficient condition for T to be *normal;* (a) For each closed subset X and each continuous function f that maps X into the closed interval [0, 1], there is a continuous function F that maps T into [0, 1] and satisfies $F(x) = f(x)$ if $x \in X$. (b) For each closed subset X and each continuous function f that maps X into the set of real numbers, there is a continuous function F that maps T into the set of real numbers and satisfied $F(x) = f(x)$ if $x \in X$.

Time Sharing. Refers to a technique in computers which facilitates a number of jobs to be run simultaneously instead of one after another. What actually happens is that each program is transferred in turn from the backing store to the main store and allowed to run for a short time, doing a spacified job. This is specially useful for the programs which are controlled by users sitting at the terminals. Thus, quite a good number of users can interact with the computer almost simultaneously.

Topological Space. Let T be a set of all subsets of a set S (including X and ϕ the null set X), will be known as topological space if it satisfies the condition

$$A \in T, B \in T \text{ then } (A \cup B) \in T, \text{ and } (A \cap B) \in T.$$

The elements of T are called open sets of the topological space X. The topology of a topological space is defined by the properties of its open sets.

Topology. The branch of *geometry concerned with the properties of an object that do not change under *homeomorphisms, i.e., when the object is bent, stretched, or shrunk but not torn or deformed so that several points on it are fused. The hole in a doughnut is such a property; for example if a rubber doughnut is distorted to the shape of a cup the hole is still there in the handle. Topology is often called rubber-sheet geometry because rubber objects can be suitably distorted. It was formerly called analysis situs. One application is in networks (e.g., an electricity—distribution network) in which the topological properties depend on the so-called *Euler characteristic (named after Leonhard Euler) V — E + F, where V is the number of vertices in the network, E the number of edges, and F the number of area enclosed by edges.

Torque symbol: T. Refers to a turning force (or moment). The torque of a force F about an axis (or point) is Fs, where s is the distance from the axis to the line of action of the force. The unit of torque is the newton metre. Note that the unit of *work,* also the newton metre, is known as the joule. Torque is not however measured in joules. The two physical quantities are not in fact the same. Work (a scalar) is the scalar product of force and displacement. Torque is the vector product and is a vector at 90° to the plane of the force and displacement.

Torr. A unit of pressure which is equal to a pressure of 101325/760 pascals (133.322 Pa).

Torsion Coefficients of a Group. If G is a commutative group with a finite set of generators, then G is a *Cartesian product* of infinite cyclic groups F_1, F_2 ...., F_m and cyclic groups H_1, H_2... H_n of finite order. The number m and the orders r_1 , r_2... r_n of H_1 H_2....., H_n form a complete system of invariants. The numbers

$r_1, r_2, \ldots. r_n$ are *torsion coefficient* of G (G is *torsion free* if n = 0).

Torsion of a Space Curve at a Point. If P is a fixed point, and P' a variable point, on a directed space curve C, Δs the length of arc on C from P to P', and Δψ the angle between the positive directions of a binormals of C at P and P' then the torsion 1/τ of C at P is defined, to within sign, by $1/\tau = \lim_{\Delta s \to 0} \pm \frac{\Delta\psi}{\Delta s}$. The sign of 1/ τ is chosen so that we have dy/ds = βτ. The torsion may be considered as a measure of the rate at which C is turning out of its osculating plane relative to the arc length s. We have

$$1/\tau = -\rho^2 \begin{vmatrix} x' & y' & z' \\ x'' & y'' & z'' \\ x''' & y''' & z''' \end{vmatrix}$$

the primes representing differentiation with respect to the arc length. The reciprocal of the torsion is the *radius of torsion*. Some authors use the symbol τ rather than 1/ τ to represent the torsion.

Torus (anchor ring). Refers to a closed curved surface having a hole in it, like a tyre inner tube. It can be generated by rotating a circle about an axis that lies in the same plane as the circle but does not cut it. A cross-section of the torus in a plane perpendicular to the axis is two concentric circles. A cross-section in any plane that contains the axis is a pair of congruent circles at equal distances on both sides of the axis. The volume of a torur is given by $4\pi dr^2$ and its surface area is given by $3\pi^2$ dr, where r is the radius of the generating circle and d is the distance of its centre from the axis. In Cartesian coordinates, a torus that has its axis lying along z-axis and a generating circle in the y-z plane with its centre at a distance d along the y-axis, has the equation:

$$\sqrt{[x^2 + y^2]} - d^2] + z^2 = r^2$$

Total Derivative. If F is a function of two variables x and y where x and y are separately some functions of a third variable t then the total derivative of F with respect to t is given, in terms of partial derivatives of F with respect to x and y as

$$\frac{dF}{dt} = \frac{\partial F}{\partial x}\frac{dx}{dt} + \frac{\partial F}{\partial y}\frac{dy}{dt}$$

This can be easily extended to the case in which F is a function of more than two variables.

Total Differential. If F is a function of two independent variables x and y, and x and y are given infinitesimal changes δx and δy respectively, then the infinitesimal change δF in F is given by

$$\delta F = \frac{\partial F}{\partial x}\,\delta x = \frac{\partial F}{\partial y}\,\delta y.$$

Tajectory. Refers to (1) the path of a moving particle of a celestial body. (2) A curve that cuts all curves (or surfaces) of a given family at the same angle. An **orthogonal trajectory** is a curve that cut all the members of a given family of curves (or surfaces) at right angles, (3) A curve or surface that fits some given law such as passing through a given set of points.

Transcendental Functions. Refers to functions which are not algebraic. The circular, exponential, and logarithmic functions all are examples of transcendental functions.

Transformation. Refers to (1) In general, any function or mapping that changes one quantity into another. (2) The changing of an algebraic expression or equation into an equivalent having a different form. For example the equation

$$(x - 3)^2 = 4x + 2$$

can be transformed into

$$x^2 - 10x - 11 = 0.$$

(3) In geometry, the changing of one shape into another by moving each point in it to a different position, generally by a specified method. For example, a plane figure may be moved in relation to two rectangular axes. Another example is when a figure is enlarged.

Transformation of Axes. This can be achieved in two ways. Axes of Coordinates are said to be *translated* when the origin of coordinates is transferred to a new position, axes remaining parallel. Axes are said to be *rotated* when these are rotate through the same angle. If the origin is changed to a point *(h, k)* then the original coordinates to a point P(x, y) become (X, Y) with respect to new axes, such that x = X + h, y = Y + k.

Transformation of Coordinates. Refers to (1) changing the position of the reference axis in a coordinate system by translation, rotation, or both, generally to simplify the equation of a curve.

(2) Changing the type of coordinate system wherein a geometrical figure is described. For example, from rectangular coordinates to polar coordinates.

Transitive, Transitive Relaltion. Refers to a relation which has the property that if A bears the relation to B and B bears the same relation to C, then A bears the relation to C. Equality in arithmetic is transitive, since if A = B and B = C, then A = C. A relation which is not transitive is **intransitive** or **nontransitive** according as there does not,or does, exist a set of objects A, B, C such that A bears the relation to B, B bears the relation to C, and A bears the relation to C. The relation of *being the father of is intransitive,* since if A is the father of B, and B is the father of C, then then A is not the father of C. The relation of *being a friend of is nontransitive,* since if A is a friend of B, and B is a friend of C, then A may or may not be a friend of C.

Translation, Translation of Axes. Means changing the coordinates of points to coordinates referred to new axes parallel to the old. Used to change the form of equations in order to aid in the study of their loci; e.g., one may desire to translate the axes so that the new origin is on the curve, which yields an equation with zero constant term, or to translate the axes until they coincide with the axes of symmetry when these are parallel to the axes, as in the case of conics, thus obtaining an equation lacking first-degree terms. **Translation formulas** are formulas expressing a translation of axes analytically. In the *plane,* these formulas are .

$$x = x' + h, y = y' + k$$

where h and k are the coordinates of the origin of the x', y' system with reference to the x, y system; i.e., when x' = y' = 0, x = h and y = k. In *space,* if the new origin has the coordinates (h, k, l) with respect to the old axes, and a point has coordinates (x', y', z') with respect to the new axes, and coordinates (x, y, z)

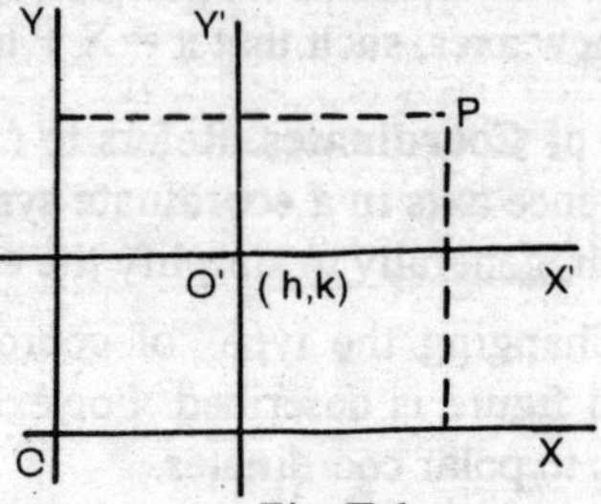

Fig. T-1.

with respect to the old axes, then $x = x' + h$, $y = y' + k$, $z = z' + l$.

Translation and Rotation. Refers to a transformation which both translates and rotates axes. Used, for instance, in the study of the general quadratic in x and y to get an equation with zero coefficients of the xy and x and y terms. The transformation formulae are

$$x = x' \cos \theta - y' \sin \theta + h,$$

$$y = x' \sin \theta + y' \cos \theta + k,$$

where h and k are the coordinates of the new origin relative to the old and θ is the angle through which the positive x-axis is rotated to be parallel to the positive x'-axis.

Translatory Motion (translation). Refers to motion involving change of position; it compares with rotatory motion (rotation) and vibratory motion (vibration). Each is associated with kinetic energy. In an object undergoing translatory motion, all the points move in parallel paths.Translatory motion is gernerally described in terms of (linear) speed or velocity, and acceleration.

Transpose. Means to move a term from one member of an equation to the other and change its sign. This is equivalent to subtracting the term from both members. The equation $x + 2 = 0$ yields $x = -2$ after transposing the 2.

Transpose of a Matrix. Means the matrix resulting from interchanging the rows and columns in the given matrix.

Transposition. Rerers to (1) the act of transposing terms from one side of an equation to the other. (2) The interchange of two objects; a cyclic permutation of two objects.

Transversal. Refers to a straight line which crosses **Coplanar lines.** Eight angles are formed when a transversal cuts two such lines as shown in the figure. These angles are paired and categorised as follows:

Alternate Angle ; (c, f), (d, e).

Corresponding Angles ; (a, e), (b, f), (c, g), (d, h).

Vertically Opposite Angles ; (a, d), (b, c), (e, h), (f, g).

Fig. T-2.

When a transversal cuts two parallel lines, all above mentioned pairs of angles are equal.

Transverse Axes

A transformation can be expressed by 2 × 2 matrix

A point (x, y) is transformed to a point (x', y') by multiplyihg the column vector of (x, y) by a matrix M

i.e. $$M\begin{pmatrix} x \\ y \end{pmatrix}$$

The transformation matrices are :

reflection in x-axis $$\begin{pmatrix} 1 & 0 \\ 0 & -1 \end{pmatrix}$$

reflection in y-axis $\begin{pmatrix} -1 & 0 \\ 0 & 1 \end{pmatrix}$

enlargement scale factor $\begin{pmatrix} k & 0 \\ 0 & k \end{pmatrix}$

k

stretch in x-direction $\begin{pmatrix} k & 0 \\ 0 & 1 \end{pmatrix}$

stretch in y-direction $\begin{pmatrix} 1 & 0 \\ 0 & k \end{pmatrix}$

rotation by α
(anticlockwise positive) $\begin{pmatrix} \cos\alpha - \sin\alpha \\ \sin\alpha -- \cos\alpha \end{pmatrix}$

shear in x-direction by k $\begin{pmatrix} 1 & k \\ 0 & 1 \end{pmatrix}$

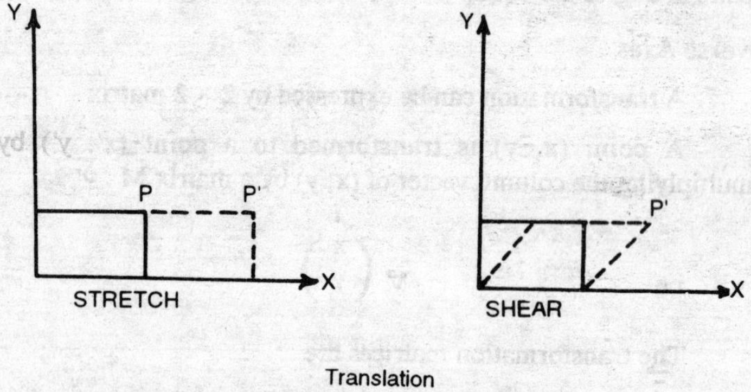

Fig. T-3.

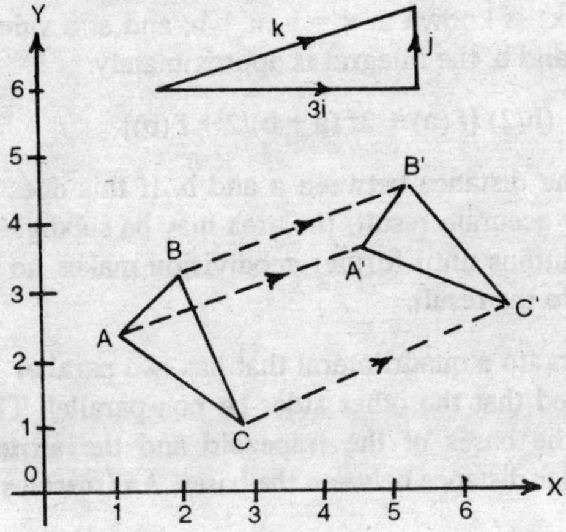

Fig. T-4.

The translation of a triangle ABC

The translation vector k = 3i + j

A translation by a in the x-direction and b in the y-direction transforms a point (x, y) to (x', y').

In matrix terms

$$\begin{pmatrix} x' \\ y' \end{pmatrix} = \begin{pmatrix} x \\ y \end{pmatrix} + \begin{pmatrix} a \\ b \end{pmatrix}$$

Transverse Wave. A wave motion wherein the motion or change is perpendicular to the direction of energy transfer. Electromagnetic waves and water waves are examples of transverse waves.

Trapezium. Refers to a plane figure having four straight sides, two of which are parallel. Its area is the product of the sum of the parallel side lengths and half the perpendicular distance between them.

Trapezium Rule. Refers to a rule used for finding the approximate area under a curve by dividing it into pairs of trapezium-shaped sections, forming vertical columns of equal width having bases lying on the horizontal axis. The trapezium rule is used as a method of numerical integration. For example, if the value of a function f(x) is known at x = a, x = b, and at a value mid-way between a and b, the integral is approximately:

$$(h/2)\ [f\,(a) + 2f\,(a + b)/2 + f\,(b)]$$

h is half the distance between a and b. If this does not give a sufficiently accurate result, the area may be subdivided into 4, 6, 8, ... columns until further subdivision makes no significant difference to the result.

Trapejoid. Refers to a quadrilateral that has two parallel sides. It is also required that the other sides be non-parallel. The parallel sides are the bases of the trapezoid and the **altitude** is the perpendicular distance between the bases. An **isosceles trapezoid**

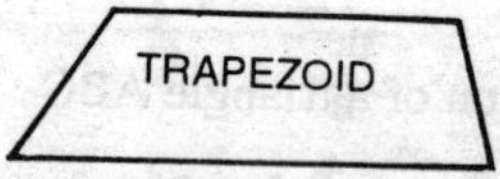

Fig. T-5.

is a trapezoid wherein the non-parallel sides are of equal length. The area of a trapezoid is equal to the product of its altitude and one-half the sum of the lengths of its bases,

$$A = h\ \frac{(b_1 + b_2)}{2}.$$

Trpezoid Rule. Refers to a formula for approximating an integral of type $\int_a^b f(x)\,dx$. The formula can be taken as the result of separating the interval from a to b in congruent subintervals by points $a = x_0, x_1, x_2..., x_n = b$, and then approximating the graph of $y = f(x)$ for x between x_k and x_{k+1} by the line segment joining the points on the graph for which x has the values x_k and x_{k+1}. The formula is

$$\frac{b - a}{n}\left[\frac{1}{2}(y_0 + y_n) + \sum_{t=1}^{n-1} y_i\right],$$

where $y_k = f(y_k)$ for each k. The numerical difference between the number given by this formula and the actual integral is not greater than

$$\frac{M(b-a)^3}{12\,n^2},$$

where M is the least upper bound of the absolute value of the second derivative of f on the interval from a to b. *Syn.* **trapezoid formula**.

Trend (Statistics). Refers to the general drift, tendency, or bent of a set of data; such, for instance, as the price of steel over a long period of tme. Particular data will generally fluctuate from the trend. Trend extending over long periods of time that is the results of slowly changing, long-lasting forces is known as **secular trend**. Trend is generally represented by a smooth mathematical function, like the trend line.

Triangle. A percussion instrument consisting of a steel rod bent into the shape of a triangle. The triangle was first used in late 18th century orchestral works to provide a "Turkish" atmosphere. musical instruments.

Triangle of Vectors. Refers to a diagrammatic represention of *composition* of vectors. If vectors $\vec{a}$ and $\vec{b}$ acting at a point O are represented in magnitude and direction by the line segments $\vec{OA}$ and $\vec{AB}$ than their resultant vector $\vec{c}$ is represented by $\vec{OB}$ along the third side of the triangle OAB. In the diagram $\vec{c}$ is the vector sum of $\vec{a}$ and $\vec{b}$ written as $\vec{c} = \vec{a} + \vec{b}$, $\vec{a}$ is the vector difference of $\vec{c}$ and $\vec{a}$ ($\vec{b} = \vec{c} - \vec{b}$) and $\vec{b}$ is the vector difference of $\vec{c}$ and $\vec{a}$ ($\vec{b} = \vec{c} - \vec{b}$). Hence, if three vectors acting at a point can be represented by the line segments $\vec{OA}$, $\vec{AB}$, $\vec{BO}$ along the sides of a triangle OAB, then their resultant will be zero and the triangle system will be in equilibrium.

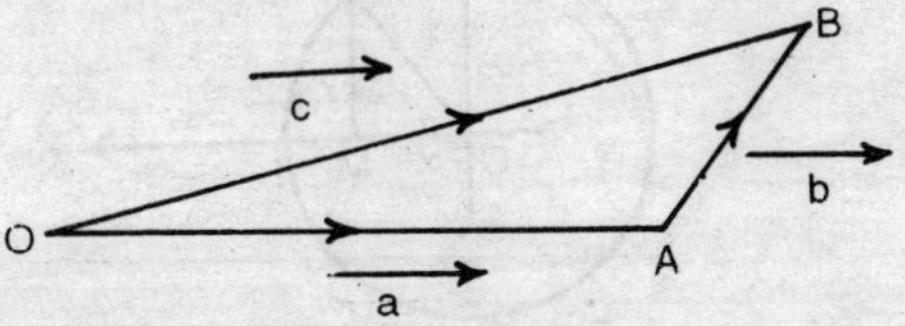

Fig. T-6.

Triangular Matrix. Refers to a square matrix wherein either all the elements above the leading diagonal or all the elements below the leading diagonal are zero. The determinant of a triangular matrix is the product of its diagonal elements.

Triangular Numbers. The set of numbers $\{1, 3, 6, 10, ...\}$ generated by triangular arrays of dots. Each array has one more row than the preceding one, the additional row having one more dot than the longest in the preceding array. The nth triangular number is $n(n+1)/2$.

Triangulation. A triangulation of a topological space T is a *homeomorphism* of T onto a polyhedron made up of the points belonging to simplexes of a simplicial complex. A **triangulable space** is a space that is homeomorphic to a simplicial complex, e.g., the surface of an ordinary sphere is triangulable since it is homeomorphic to the surface of an inscribed regular tetrahedron, the homeomorphism made up of projections of points of the sphere onto the teterahedron along radii (and of points of the tetrahedron onto the sphere along radii). The surface of a regular teterahedron is a simplicial complex whose simplexes are triangles. This mapping of the tetrahedron onto the sphere separates the sphere into four spherical triangles corresponding to the four faces of the tetrahedron.

Trichotomy, Trichotomy Property. The property sometimes considered for an ordering that exactly one of the statements $x < y$, $x = y$ and $y < x$ is true for any x and y.

Trigonometric Functions. *Syn. Circular Functions*. Functions of angles have been defined in terms of trigonometric ratios. If a point P(x, y) lies on the circumference of a unit circle $x^2 + y^2 = 1$; with centre O, the origin, the trigonometric functions of the angle θ made by OP with positive direction of x-axis, will be given as follows:

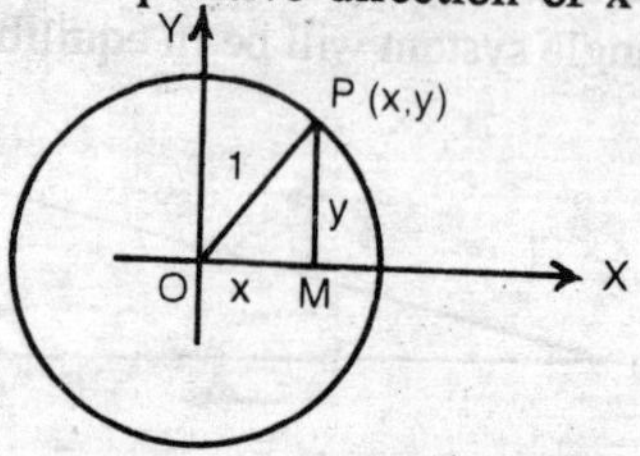

Fig. T-7.

$\sin\theta = y$, $\cos\theta = x$, $\tan\theta = \frac{y}{x}$, $\cot\theta = \frac{x}{y}$, $\sec\theta = \frac{1}{x}$,

$\text{cosec}\,\theta = \frac{1}{y}$.

Trigonometric Identities. Refers to identities which involve trigonometric functions. For example,

$\sin^2\theta + \cos^2\theta = 2$, $\sec^2\theta = 1 + \tan^2\theta$, $\text{cosec}^2\theta = 1 + \cot^2\theta$ etc.

Trigonometric Series. A series involving trigonometric functions, expressed as

$$\tfrac{1}{2} a_0 + (a_1 \cos x + b_1 \sin x) + (a_2 \cos 2x + b_2 \sin 2x) + \ldots\ldots$$

Fourior Series is the most important trigonometric series.

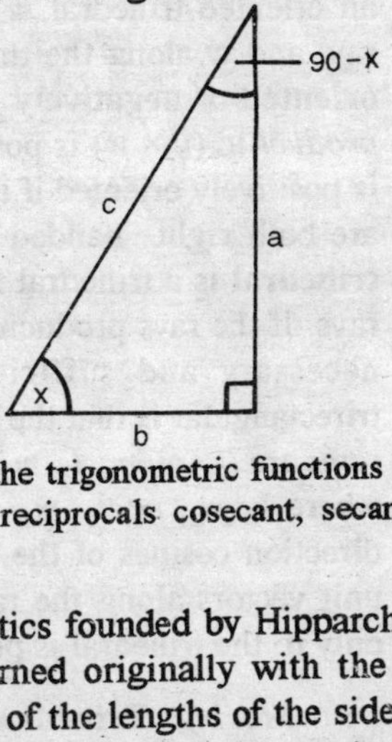

Fig. T-8.

TRIGONOMETRY: Definitions of the trigonometric functions since, consine, and tangent and of their reciprocals cosecant, secant and cotangent.

Trigonometry. A branch of mathematics founded by Hipparchus in the mid-2nd century BC, concerned originally with the measurement of triangles. The ratios of the lengths of the sides of a right-angled triangles are used to define the sine, cosine, and tangent of one of the angles of the triangle. Trigonometry deals with the properties of these and related functions. Its study is essential to most branches of physical and mathematics, particularly those involving cyclic quantities.

Trihedral. Refers to (1) a figure obtained from three noncoplanar lines which intersect in a point. A trihedral obtained by three directed lines is a **directed trihedral**. The trihedral produced by the three axes of a system of Cartesian coordinates in space is a directed trihedral, known as the **coordinate trihedral.** (2) The union of three noncoplanar rays having the same initial point. An **oriented trihedral** is a trihedral for which one of the rays is designated as the first, another as the second, and the remaining one as the third. An oriented trihedral is **left-handed** if, when the thumb of the left hand extends from the initial point along the first ray, the fingers fold in the direction in which the second ray could be rotated through an angle of less than 180° to concide with the third; it is **right-handed** if, when the thumb of the right hand extends from the initial point along the first ray, the fingers fold in the direction in which the second ray could be rotated through an angle of less than 180° to coincide with the third. If vectors, *u*, *v* and *w* are represented by arrows having initial points at the initial points of the rays of an oriented trihedral, *u* along the first ray, *v* along the second ray, and w along the third ray, then the trihedral is **positively oriented** or **negatively oriented** according as the *triple scalar product* $u. (u \times w)$ is positive or negative. An oriented trihedral is positively oriented if an only if it and the coordinate trihedral are both right handed or both left-handed. A **trirectangular trihedral** is a trihedral formed by three mutually perpendicular rays. If the rays producing an oriented trihedral are L_1, L_2, L_3, a necessary and sufficient condition that the trihedral be trirectangular is that the absolute value of the determinant whose rows are l_1, m_1 n_1; l_2, m_2 n_2; and l_3, m_3, n_3 (in these orders) be 1, where l_1, m_1, n_1; l_2, m_2, n_2; and l_3, m_3, n_3 are, respectively, the direction cosines of the rays L_1, L_2 and L_3, i.e., components of unit vectors along the rays. The determinant is positive if and only if the trihedral is positively oriented.

Moving **Trihedral of Space Curves and Surfaces.** The *moving trihedral* (or *trihedral*) of a directed space curve C is the configuration made up of the tangent, principal normal, and binormal of C at a variable point of C. The moving trihedral of a **surface** relative to a directed curve on the surface is defined as follows: For a point P of a directed curve C on a surface S, let α be the unit vector from P in the positive direction of the tangent to C at P, let γ by the unit vector from P in the positive direction of the normal to S at P, and let β be the unit vector from P in the tangent plane to S at P, and such that α, β, γ have the same mutual orientation as the x, y, z axes. Then axes directed along the vectors α, β, γ form the moving trihedral of S relative to C. The rotations of the moving **trihedral** of a surface are a certain set of six particular functions determining the orientation of the *moving trihedral* of the surface, but not its position in space.

Trihedral Angle. Refers to a polyhedral angle with three faces. Two trihedral angles are symmetric if they have their face angles equal in pairs but arranged in opposite order. Such trihedral angles are not superposable.

Trillion. (1) In the U.S. and France, the number represented by one followed by 12 zeros or 10^{12}. (2) In England, the number represented by one followed by 18 zeros or 10^{18}.

Trinomial. Refers to an algebraic expression having three variables in it. For example, $2x + 2y + z$ and $3a + b = c$ are trinomials.

Triple Integral. Refers to the result of integrating the same function three times. For example, if a function f(x, y, z) is integrated first with respect to x, holding y, and z constant, and the result is then integrated with respect to y, holding x and z constant, and the resultant double integral is then integrated with respect to z holding x and y constant, the triple integral is

$$\iiint f(x, y, z)\, dz\, dy\, dx.$$

Triple Scalar Product. A product of three vectors, the result of which is a scalar quantity. For vectors $\vec{a}, \vec{b}, \vec{c}$ it is defined as $\vec{a}.(\vec{b} \times \vec{c})$. It is denoted by $[\vec{a}\,\vec{b}\,\vec{c}\,]$ and is also known as *box* product. Geometrically it is equal to the volume of a parallelo piped with $\vec{a}, \vec{b}, \vec{c}$ forming its coterminus edges. The box product is zero if the three vectors are coplanar.

Triple Vector Product. Refers to a product of three vectors, the result of which is a vector. It is a vector product of two vectors, one of which is itself a vector product. That is:

$$A \times (B \times C) = (A.C)\,B — (A.B)C.$$

Similarly $(A \times B) \times C = (A.C)B - (B.C)A$.

These are equal only when A, B and C are mutually perpendicular.

Trirectangular. Means having three right angles, e.g., a spherical triangle can have three right angles; such a triangle is trirectangular. A trirectangular trihedral angle is a trihedral angle whose three dihedral angles are right angles.

Trisection of an Angle. Refers to the problem of trisecting any angle with straight edge and compasses, alone. It was proved impossible by P.L. Wantzel in 1847. Any angle can be trisected, however, in several ways, for instance, by the use of a *protractor*, the *limacon* of Pascal, the *conchoid* of Nicodemes, or the *trisectrix* of Maclaurin.

Trisectrix. The locus of the equation $x^3 + xy^2 + ay^2 3ax^2 = 0$. The curve is symmetric with respect to the x-axis, passes through the origin and has the line $x = - a$ as an asymptote. It is of interest in connection with the problem of trisecting a given angle, for if a line having an angle of inclination 3A is drawn through the point (2a, 0), then the line passing through the origin and the point of intersection of this line with the trisectrix has an angle of inclination A. Also known as the trisectrix of Maclaurin.

Trivial, Trivial Solutions of a Set of Homogeneous Linear Equations. Zero values for all the variables, *trivial* as they are a solution of any system of homogeneous equations. Solution wherein at least one of the variables has a value different from zero are **nontrivial solutions.**

Trivial Solution. A solution to an equation or set of equations which is obvious and gives no useful information about the relationships between the variables involved. For example, $x^2 + y^2 = 2x + 4y$ has a trivial solution $x = 0$; $y = 0$.

Trichoid. Refers to the plane locus of a point on the radius of a circle, or on the radius produced, as the circle rolls (in a plane) on a fixed straight line. If a is the radius of the rolling circle, b the distance from the centre of the circle to the point describing the curve, and θ the angle (in radians) subtended by the arc that has contacted the fixed line in getting to the point under consideration, then the parametric equations of the *trochoid* are

$$x = a\theta - b \sin \theta \text{ and } y = a - b \cos \theta.$$

when b is greater than a, the curve has a loop between every two arches, nodes at $\theta = \theta_1 + n\pi$, where $0 < \theta_1 < \pi$ and $a\,\theta_1 - b \sin \theta_2 = 0$; it is then called a **prolate cycloid**. If b is less than a, the curve never touches the base line; it is then called a **curtate cycloid** (the names prolate and curtate are sometimes interchanged). As b approaches zero, the curve tends to smooth out nearer to the straight line described by the centre of the circle. When b = a the curve is known as a **cycloid.**

Truncated. Refers to a solid generated from a given solid by two non-parallel planes cutting the given solid.

Truth Table. In logic, a mechancial procedure (sometimes known as a matrix) which can be used to define certain logical operations, and to find the truth value of complex propositions or statements having combinations of simpler ones. A truth table lists, in rows, all the possible combinations of truth values (T = 'true', F = 'False') of a proposition or statement and given an initial assignment or truth of falsity to the constituent parts, mechanically assigns a value of the whole. The truth table definitions for cojunction, disjunction, negation, and implication are shown at those headwords.

An example of a truth table for a complex, or compound proposition is shown in illustration.The assigning of values proceeds in this way: on the basis of the truth values of P and Q, the simple propositions are given truth values, written under the signs ($\wedge$ in $P \wedge Q$, ~ in ~ P). Using these, truth values can then be assigned to the whole; in the example, this is in effect a complex disjunction and the values are written under the sign V.

Thus in the case where P is true and Q is false, $P \wedge Q$ will be false, ~P will be false, and therefore the whole will be false.

PQ	(P ∧ Q)	V	~P
TT	T	T	F
TF	F	F	F
FT	F	T	T
FF	F	T	T

An example of a truth table.

Truth Value. Refers to the truth or falsity of a proposition in logic. A true statement or proposition is shown by T and a false one by F. In computer logic, the digits 1 and 0 are used to denote truth value.

Turing Machine. A hypothetical computing machine used to determine whether a particular mathematical problem can be solved by a computation procedure (algorithm). It is postulated that the machine has an infinite tape that stores character in a number of discrete locations. The machine goes through a procedure of scanning and altering the characters, regarded as a series of active states. If the problem is soluble, the machine settles in a passive state, in which the tape contains the solution. The concept was developed in 1936 by Alan Mathison Turing (1912-54) and others and represented an important advance in computer logic.

Turning Point. Refers to a point on the graph of a function where the slope (gradient) of the tangent to a continuous curve changes its sign. If the slope changes from positive to negative, that is, the y-coordinate stops increasing and starts decreasing, it is maximum point. If the slope changes from negative to positive it is a minimum point. Turning points may be local maxima and minima or absolute maxima and minima. All turning points are stationary points . At a turning point the derivative, dy/dx of the curve y = f(x) is zero.

Two-Dimensional. Having length and breadth but no depth. Flat shapes, like circles, squares, and ellipses, can be described in a coordinate system making use of only two variables, for example, two-dimensional Cartesian coordinates with an x-axis and a y-axis.

U

Undetermined, Undetermined Coefficients. Unknowns to be determined in order to satisfy certain conditions, e.g., if it is desired to factorise $x^2 - 3x + 2$, the factors may be taken to be $x + a$ and $x + b$, where a and b are to be determined in order to make the product of these two factors equal to the original expression, i.e.,

$$x^2 + (a + b)\,x + ab \equiv x^2 - 3x + 2,$$

whence $a + b = -3$ and $ab = 2$, from which $a = -1. b = -2$, or $a = -2, b = -1$. The **method of undetermined coefficients** used for differential equations is shown by the following examples: To find a particular solution of the equation $y'' + 2y' - 5y = 5 \sin x$, substitute $y = A \sin x + B \cos x$, to get $A = -3/4$; $B = -1/4$ to find a particular solution of $y'' - 3y' + 2y = 27x^2e^x$ substitute

$$y = x^2\,(Ax^2e^2 + Bxe^x + Ce^x),$$

to get $A = 3/4$, $B = -1$, $C = 1$ [the factor x^3 was introduced because $y'' - 3y' + 2y = (D + 2)(D - 1)^2\, y$ and e^x is a solution of $(D - 1)\,y = 0$; the terms in $Ax^2e^2 + Bxe^x + Ce^x$ represent all types obtainable by successive differentiation of $27x^2 e^x$].

Unicursal, Unicursal Curve. Refers to curve that has parametric equations $x = \theta(t)$, $y = \phi(t)$ for which θ and ϕ are rational functions of t.

Uniform Acceleration. Refers to acceleration which remains constant throughout the motion of an object.

Uniform Convergence. The functional series

$$u_1(x) + u_2(x) + \ldots\ldots + u_n(x) + \ldots\ldots,$$

is known as uniformly convergent on the interval (a, b) if for any small arbitrary $\varepsilon > 0$, there exists an integer N such that for all $n \geq N$, $| s(x) - s_n(x) | < \varepsilon$, for *any value of* x in (a, b), where s(x) and $s_n(x)$ denote the sum of the infinite series and sum to n terms of the series respectively.

Unique. Leading to one and only one result; made up of one, and only one. The product of two integers is unique; the square root of an integer is not unless the integer is 0.

Unique Solution. Refers to the only possible value of a variable that can satisfy an equation. For example, $x + 2 = 4$ has the unique solution $x = 2$, but $x^2 = 4$ has no unique solution because $x = +2$ and $x = -2$ both satisfy the equation.

Unit. Refers to a reference value of a quantity that is used to express other values of the same quantity.

Unit Circle. Refers to a circle having radius unity. Its area is π and circumference is 2π.

Unit Cube. Refers to a cube having each edge of unit length. Its volume is taken as unit volume.

United States Rule. (Mathematics of Finance). A rule for evaluating a debt upon which partial payments have been made: Apply each payment first to the interest, any surplus being deducted from the principal. If a payment is less than all interest due, the balance of the interest must not be added to the principal and cannot draw interest. This is the legal rule.

Unit Matrix. (Identify matrix) Symbol : I. Refers to a square matrix wherein the elements in the leading diagonal are all equal to one, and the other elements are zero. If a matrix A with m rows and n columns is multiplied by an $n \times n$ unit matrix I, it remains unchanged, that is $IA = A$. The unit matrix is the *identity matrix* for matrix multiplication.

Unit Sphere. A sphere of unit radius. The equation.

$$x^2 + y^2 + z^2 = 1,$$

represents a sphere of unit radius with centre at the origin.

Unit Vector. Refers to a vector having a magnitude of one unit. Any vector r can be expressed in terms of its magnitude, the scalar quantity r, and the unit vector r' which has the direction as rr = rr'. In three-dimensional Cartesian coordinates with origin 0 unit vectors i, j and k are used in the x–, y–, and z – directions respectively.

Unity. The symbol 1. It represents an abstract number of magnitude 1. In Boolean algebra, theory of groups, rings and fields, it stands as symbol of unit element not necessarily meaning the absolute number 1.

Univac. Refers to an automatic digital computing machine manufactured by the Sperry Rand Corporation. Univac is an acronym for *Universal Automatic Computer.*

Unknown, Unknown Quantity. (1) Refers to a letter of literal expression whose numerical value is implicit in certain given conditions by means of which this value is to be determined. It is used mostly in connection with equations. In the equation $x + 2 = 4x + 5$, x is the unknown. (2) More properly, a symbol as described in (2) is a **variable**, and "unknown" refers to a member of the solution set. Thus, for the equation $x^2 - 5x + 6 = 0$, x is a variable and the unknowns are 2 and 3, the values of x that make the equation a true statement.

Unlike Terms. Refers to terms which are not similar (like). These terms can't be reduced to a single term.

Unstable Equilibrium. Refers to an equilibrium which is such that if it is disturbed a little there is a tendency for the system to move away from its original position rather than to come back to original position of equilibrium.

Upthrust. Refers to an upward force on an object in a flud. In a fluid in a gravitational field the pressure increases with depth. The pressures at different points on the object will thus differ and the resultant is vertically upward.

Utility Programs. Refers to programs that help in the general running of a computer system. They can be used, for example, to make copies of files (organized collections of data) and to transfer data from one storage device to another, as from a magnetic tape unit to a disk store.

Urysohn's Lemma. If P and Q are two nonintersecting closed sets (in a normal topological space T), there exists a real function f defined and continuous in T and such that $0 \leq f(p) \leq 1$ for all p, with $f(p) = 0$ for p in P, and $f(p) = 1$ for p in Q.

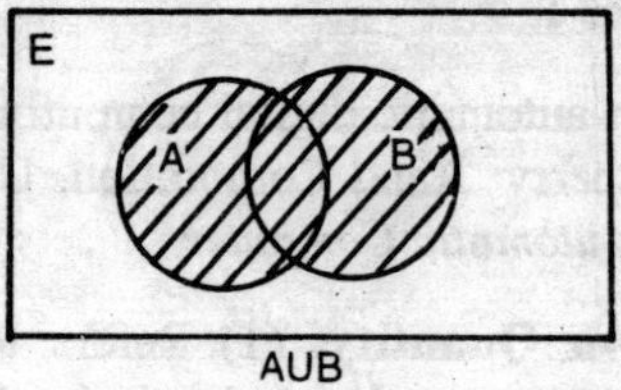

Fig. U-1.

The shaded area in the Venn diagram is the union of sets A and B.

V

Validity. Refers to a property of arguments sufficiently supported by facts or reasons. An argument is valid if it is impossible to prove that the conclusion be false.

Value, Absolute, Value. See Absolute — absolute value of a real number, Modulus — modulus of a complex number, vector—absolute, value to a vector.

Variable. A mathematical symbol for a quantity that can take any value from a set of values called the range. A variable that can take any value between two given values is called continuous; otherwise it is discrete. A quantity that can only take one value is called a constant.

Variance. In a set of numbers, usually measurements, the quantity obtained by summing the squares of the differences between each number and the average value of the set and then dividing by the number of members of the set. Variance and, more often, its square root, called the **standard deviation,** are used to estimate scatter or random error in experimental results.

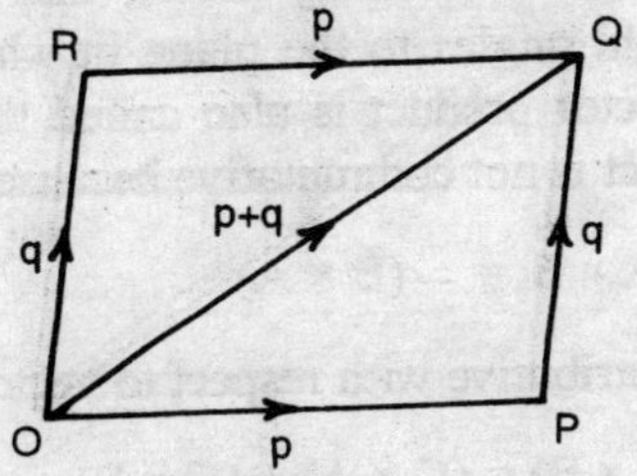

Fig. V-1.

VECTORS: The two vectors OP (p) and PQ (q) add to give the resultant vector OQ (p + q). OR = PQ = q; OP = RQ = p. OQP is a vector triangle and ORQP is a parallelogram of vectors.

Vector. A quantity that has both magnitude and direction. Examples of vectors included velocity, force, magnetic flux density, etc. A vector needs three numbers (called components) to be defined, each number representing its magnitude in one of three mutually perpendicular directions. Two vectors are added by adding the corresponding components and may be multiplied to give either a *scalar quantity or an other vector.

Vector Difference. Refers to difference of vectors $\vec{A}$ and $\vec{B}$ represented by $\vec{A}$—$\vec{B}$ is equivalent to the sum of $\vec{A}$ and $\vec{B}$.

Vector Multiplication. Refers to multiplication of two or more vectors. This can be defined in two ways according to whether the result is a vector or a scalar.

Vector Product. Refers to a multiplication of two vectors so as to give a vector. The vector product of A and B is written A × B. It is a vector of magnitude A B sin θ, where A and B are the magnitudes of A and B and θ is the angle between A and B. The direction of the vector product is at right angles to A and B. It points in the direction in which a right-handed screw would move turning from A towards B. An example of a vector product is the force F on a moving charge Q in a field B with velocity v (as in the motor effect). Here

$$F = QB \times v$$

Another example is the product of a force and a distance to give a moment (turning effect), that can be represented a vector at right angles to the plane in which the turning effect acts. The vector product is also called the *cross product*. The vector product is not commutative because

$$A \times B = -(B \times A)$$

It is distributive with respect to vector addition

$$C \times (A \times B) = (C \times A) + (C \times B)$$

The magnitude of A × B is equal to the area of the parallelogram of which A and B form two non-parallel sides. In a three-dimensional Cartesian coordinate system with unit vector, i, j and k in the x, y and z directions, respectively,

$$A \times B = (a_1 i + a_2 j + a_3 k) \times (b_1 i + b_2 j + b_3 k)$$

This can be written as determinant.

Vector Projection. Refers to the vector resulting from an orthogonal projection of one vector on another, For example, the vector projection of A on B is b A cos θ where θ is the smaller angle between A and B, and b is the unit vector in the direction of B.

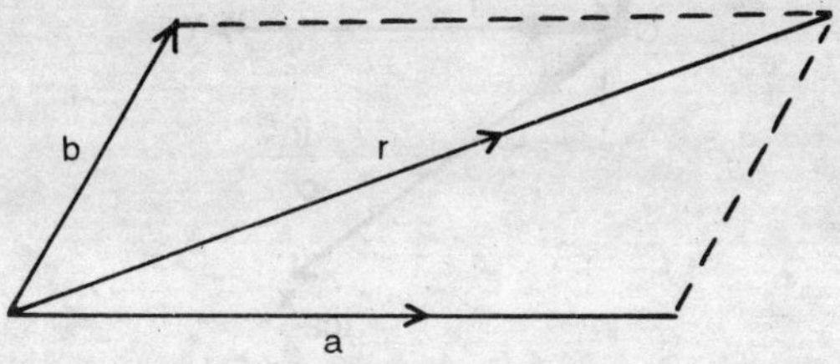

Fig. U-2.

The parallelogram law r is the resultant of a and b

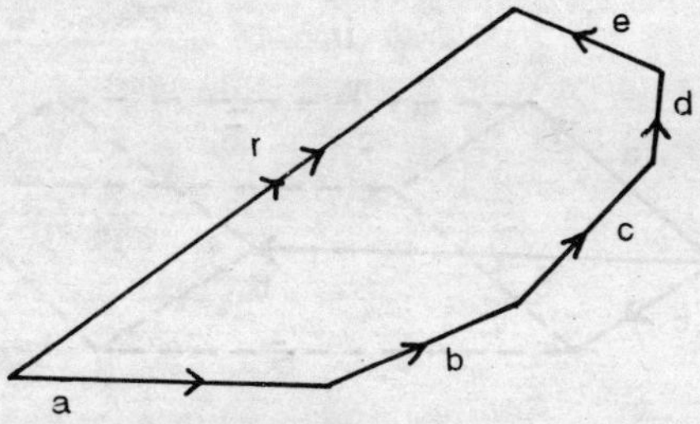

Fig. U-3.

The polygon of vectors r is the resultant

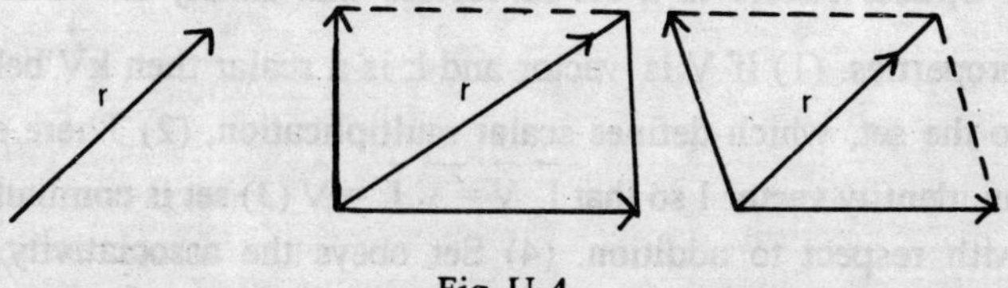

Fig. U-4.

Resolution of the vector r into different pairs of componenents

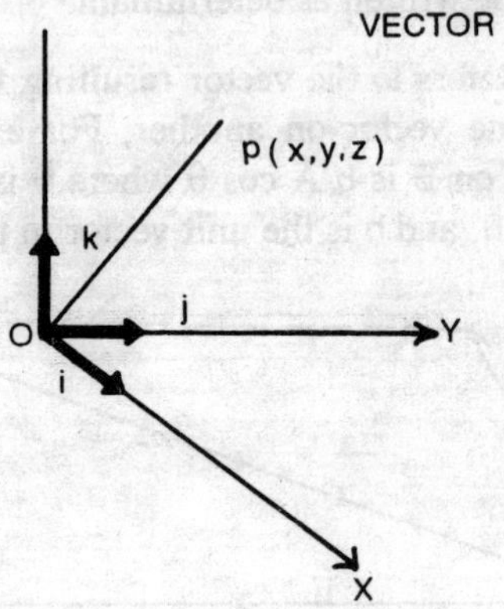

Fig. V-5.

Basic vectors. The vector OP can be represented as ix + jy + kz

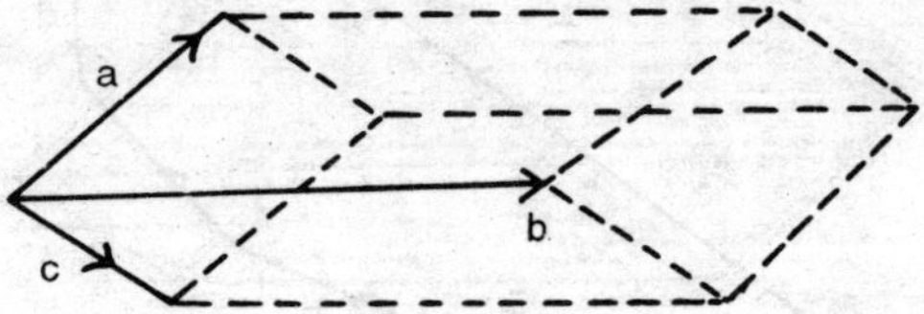

Fig. V-6.

Vector product c = a × b

Vector Space. Refers to a set of vectors that satisfy the following properties. (1) If $\vec{V}$ is vector and k is a scalar then $k\vec{V}$ belongs to the set, which defines scalar multiplication, (2) There exists an identity vector I so that $\vec{I}.\ \vec{V} = \vec{V}.\vec{I}. = \vec{V}$ (3) set is commutative with respect to addition. (4) Set obeys the associativity with respect to addition as well as multiplication. (5) Set obeys distributive property of vectors; such a set of vectors is called Vector Space. The vector space may be two dimensional, three dimensional,....., or n dimensional.

Vector Sum. Refers to the result of adding two vectors $\vec{A}$ and $\vec{B}$ if components of $\vec{A}$ and $\vec{B}$ are (a_1, a_2, a_3) and (b_1, b_2, b_3) respectively then the sum $\vec{A} + \vec{B}$ will have components $(a_1 + b_1, a_2 + b_2, a_3 + b_3)$. It can be achieved diagrammatically too by drawing the two vectors head to tail. The sum is the vector represented by the line segment from the tail of the first to the head of the second.

Velocity. It is the rate of change of displacement with respect to time for motion in a straight line. Velocity is a vector quantity while speed is its magnitude. Velocity of a point with displacement x is dx/dt at any time t, in case of rectilinear motion. When the displacement is curvilinear that is, along a curve, the velocity is in the direction of the tangent to the curve at the point.

Velocity of Light. Generally denoted by c. It is approximately 3×10^5 km/sec.

Vertex. Refer to (1) a point where lines or planes meet in a figure; for example, the top point of a cone or pyramid, or a corner of a polygon or pholyhedron.

(2) One of the two points where an axis of a conic cuts the conic.

Vertical Line. Refers to a line in a direction towards or away from the centre of the earth.

Vertically Opposite Angles. One of the two pairs of equal angles when two straight lines interesect each other.

Vibration (Oscillation). Refers to any regularly repeated to-and-fro motion or change. For example, the swing of a pendulum, the vibration of a sound source and the change with time of the electric and magnetic fields in an electromagnetic wave.

Virtual Work. Refers to the work done when a system is displaced infinitesimally from its position. The virtual work is zero if the system is in equilibrium.

Visual Display Unit (VDU). Refers to a computer terminal with which the user can communicate with the computer by mean of a keyboard similar to that of a typewriter, this input and also the output from the computer appears on a television screen. A VDU can operate both as an input and output device. The information displayed is in the form of words, numbers, etc. A *graphic display* is a similar device wherein information appears as graphs or other drawings as well as text.

Vital Statistics. Refers to statistics dealing with the data on death, birth, marriages etc.

Volt. Refers to unit of electromotive force (E.M.F.) and of potential difference. Unit volt is the potential difference between two points in a circuit through which one ampere current flows.

Volume. It is a measure of the space that is occupied by a solid or bounded by a closed surface. For example, the volume of a parallelopiped is the product of its length, breadth and height.

W

Waring's Problem. To prove the assertion made by Waring in 1770 that, for any integer n, there is a least integer g(n) such that any integer can be shown as the sum of not more than g(n) numbers, each of which is an nth power of an integer, this problem was solved by Hilbert (1909). Lagrange showed in 1770 that any integer can be represented as the sum of not more than 4 squares. It is now known that $g(3) = 9$, $19 \leq g(0) \leq 30$, and $g(n) = 2^n + A—2$, except for a finite (possible empty) set whose members are either 4 or numbers greater than 50,000, where, A is the greatest integer less than $(3/2)^n$, e.g., $g(5) = 37$ and $g(6) = 73$.

Watt. A metric unit used to measure power; the power required to keep a current of one ampere flowing under a potential drop of one volt; 1/736 of one horsepower (English and American).The international watt is defined terms of the *international ampere* and the *international volt* and differs slightly from the **absolute watt,** which is equivalent to 10^7 ergs (one joule) of work per second.

Wave, Wave Equation. The partial differential equation

$$\frac{\partial^2\psi}{\partial x^2} + \frac{\partial^2\psi}{\partial y^2} + \frac{\partial^2\psi}{\partial z^2} = \frac{1}{c^2}\frac{\partial^2\psi}{\partial t^2}.$$

In the theory of sound, this equation is satisfied by the velocity potential (in a perfect gas); in the theory of elastic vibrations, it is satisfied by each component of the displacement; and in the theory of electric or electromagnetic waves, it is satisfied by each component of the electric or magnetic (force) vector. The constant e is known as the velocity of propagation of the periodic disturbance.

Wave Equation. Refers to the partial differential equation

$$\frac{\partial^2 y}{\partial t^2} = c^2\frac{\partial^2 y}{\partial x^2}$$

Wavefront. Refers to a continuous surface associated with a wave radiation, wherein all the vibrations concerned are in phase. A parallel beam has plane wavefronts: the output of a point source has spherical wavefronts.

Wavelength. Symbol : λ. The distance between the ends of one complete cycle of a wave. Wavelength relates to the wave speed (c) and its frequency (v) thus:

$$c = v\lambda.$$

Wave Number. Symbol: σ. It is the reciprocal of the wavelength of a wave. It is the number of wave cycles in unit distance, and is used in spectroscopy. The unit is the metre^{-1} (m^{-1}). The circular wave number (symbol: k) is given by

$$k = 2\pi\sigma.$$

Wearing, Wearing Value of Equipment. The difference between the purchase price and the scrap value. Also known as **original wearing value**. The difference between the book value and scrap value is the **remaining wearing value**.

Weber. Symbol: Wb. Refers to the SI unit of magnetic flux and is equal to the magnetic flux that, linking a circuit of one turn, produces an e.m.f. of one volt when reduced to zero at a uniform rate in one second. 1 Wb = 1 Vs.

Weddle's Rule. According to this rule,

$$\int_a^b f(x)\,dx = \frac{3h}{10}\left[y_0 + 5y_1 + y_2 + 6y_3 + y_4 + 5y + 2y_6 + 5y_7 + y_8 + \ldots\ldots\right]$$

where $h = \dfrac{b - a}{n}$ $y_0, y_1, y_2, \ldots..$ are the value of f (x) at x = a, a + h, a + 2h etc.

Weight. It is the force of attraction of the earth on any object near its surface. It is given by w = mg, where w is the weight, m the mass of the object and g the acceleration due to gravity. Due to change in values of g from place to place, the weight also changes, whereas the mass remains constant.

Weighted Average (mean). It is the mean of a set of numbers, each number having been associated with weights in the form of relative numerical importance. For example, if a student obtains 68, 83, 95 marks in the subjects English, Science and Mathematics respectively, and if the relative importance, of these subjects are in the ratio of 2 : 3 : 4, known as the weights, the weighted average (mean) will be

$$(2 \times 68 + 3 \times 83 + 4 \times 95)/(2 + 3 + 4) = 85 \text{ marks.}$$

Weightlessness. Refers to an apparent loss of weight experienced by an object in free fall. Thus for a person in an orbiting spacecraft, the weight in the Earth's frame of reference is the centripetal force necessary to maintain the circular orbit. In the frame of reference of the spacecraft, the person feels that he has no weight.

Whole Numbers. Symbol: W. The set of integers {1, 2, 3,...}, excluding zero.

Width. For a convex set in a plane, the greatest lower bound of numbers w such that the set is between two parallel lines separated by distance w. The same definition is given for convex sets in n dimensional space, except that "parallel lines" is replaced by "parallel hyperplanes". However, there are other usages: e.g., a box with edges $a < b < c$ is said to have width b and length c, as "a sheet of steel 24 cm long, 4 cm wide, and 1/2 cm thick."

Weierstras's Theorem. The theorem states that if a sequence of holomorphic functions converges uniformly in a domain then the limit function will also be holomorphic in the same domain.

Well Ordered Set. A partially ordered set is considered to be well ordered if every subset of it contains first element. For example, the set of natural numbers with natural order is well-ordered.

Wiener Process. A *stochastic process* {X(t)}: t 7 > 0 is a **Weiner process** (or **Brownian-motion process**) if (i) X (0) = 0; (ii) for each t, X (t) is a normal random variable with zero mean; (iii) if $a < b < c < d$, the random variables X (b) – X (a) and X(d) – X(c) are independent and have the same distribution whenever b—a = d—c. Each Wiener process is a *Martingale* and has a constant B such that if $0 \leq t_1 \leq t_2$, than $X(t_2)—X(t_1)$ is *normal* with mean 0 and variance B $(t_2—t_1)$. Weiner processes have applications to the study of Brownian motion, price movements in stock markets, quantum mechanics, etc.

Wilson's Theorem. The necessary and sufficient condition for the number (n—1) ! + 1 to be divisible by n is that n is a prime number. Its necessary part was proved by Leibnitz while the sufficient part was proved by Lagrange. The number (n—1) ! + 1 is called *Wilson* **number**.

Winding, Winding Number. The number of times a closed curve in the plane passes around a designated point of the plane in the counterclockwise (positive) direction. *Tech.* Let C be a closed curve in the plane that is the image of a circle for a continuous transformation. This means that C has parametric equations of type x = u(t), y = v(t) for $0 \leq t \leq 1$, where u and v are continuous functions with u (0) = u (1) and v (0) = v (1); or (equivalently) an equation of type w = f(z) for | z | =1, where z and w are complex numbers and f is a continuous function. If P is chosen as a point that is not on C, the numbers $\{t_i\}$ satisfy $t_o = 0 < t_2 < t_1 <...< t_n = 1$ and Q_i is the point for which the parameter t is given by t_i (i = 1, 2,..., n), then there is a positive number E such that the quantity $\frac{1}{2\pi}\sum_{1}^{n} \theta_i$ has a value n(C, P) that is independent of the choice of the numbers $\{t_1\}$ provided $(t_i — t_{i-1}) < E$ for each i, where θ_i is the angle (in radian measure) from the line PQ_{i-1} to the line PQ_o for each i. This number n(C, P) is an integer and is the **winding number** of C relative to P, or the **index** of P relative to C. The winding number of a curve remains unchanged by a continuous deformation of the curve for which the curve does not pass through the point P, e.g., if p(z) is a polynomial of degree n with $p(0) \neq 0$, and C_k is the image of the circle | z | = K for the mapping w = p(z) , then for sufficiently large K the winding number of C_k relative to the origin is n, and for sufficiently small K the winding number of C_k relative to the origin is 0. Since these curves can be continuously deformed into each other (by letting K change continuously), there must be a value of K for which the curve C_k passes through the origin and value of z for which p(z) = 0. This gives a proof of the *fundamental theorem of algebra*. If P is the complex number a and the curve C is defined by w = f(z), where f is the *piecewise differentiable*, then

$$n(C, a) = \frac{1}{2\pi i} \int_c \frac{dz}{z - a}.$$

Witch. Refers to a plane cubic curve which is defined by drawing a circle of radius a, tangent to the x-axis at the origin, then drawing a line through the origin and forming a right triangle having its hypotenuse on this line and one leg parallel to the x-axis, the other parallel to the y-axis, and passing respectively, through the points of intersection of this line with the circle and

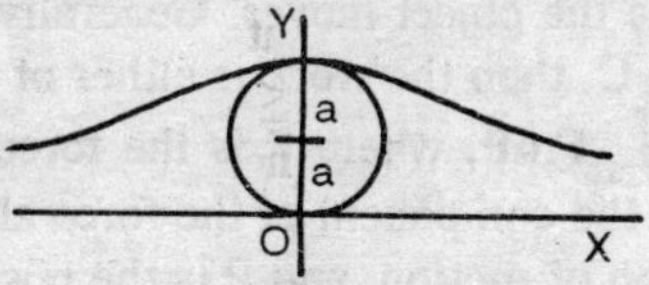

Fig. W-1.

the line y = 2a. Which is then the locus of the intersection of the legs of all such triangles. Its equation in rectangular coordinates is given by $x^2y = 4a^2(2a - y)$. The *witch* is generally known as the **witch of Agnesi,** after Donna Maria Gaetana Agnesi, who discussed the curve.

Witch of Agnesi. This is a curve whose cartesian equation is $4a^2(2a - y) = x^2y$. Its parametric form .

$$x = 2a \cot \theta, \; y = 2a \sin^2 \theta$$

is sometimes more convenient. For the shape of the curve, take two parallel lines at a distance 2a apart. Draw a circle that touches these lines. Let one point of contact be taken as origin of coordinates, O, the tangent at O as the x-axis and the diameter through O as y-axis. Draw any line through O cutting the circle at A and the parallel tangent at B. Draw a line through A parallel to x-axis and a line through B parallel to y-axis and let

P be their point of intersection. The locus of P, i.e., totality of all possible points P is the required curve known as Witch of Agnesi given by Maria Agnesi in 1748. One important property of this curve is that the area between the curve and the x-axis is four times that of the generating circle.

Work. For a constant force acting on an object moving in the direction of the force, the **work** done on the object by the force is given by the product of the magnitude of the force and the distance the object moves, e.g., the work necessary to lift a 25 pound bag of flour from the floor to the top of a 3 foot counter is 75 foot-pounds. When the force is constant but at an angle θ with the line along which the object moves, then the work W is (F cos θ) s, where F is the magnitude of the force, F cos θ is the magnitude of the component of the force in the direction of motion, and s is the distance the object moves. Generally, if the object moves along a curve C, then the work is either of the line integrals, $W = \int_c F_t\, ds = \int_c \mathbf{F.dP}$, where **F** is the force (a vector), F_t is the magnitude of the component of the force along the tangent to C in the direction of motion, and P is the position vector of points on C. If C is described by parametric equations using time *t* as a parameter, than $W = \int_c \mathbf{F.V.}\, dt$, where F. V. is the scalar product of **F** and the velocity vector **V** = d **P**/dt. The total work done on a particle by all forces acting on it is equal to the change in kinetic energy.

Wronskian. This is a determinant obtained while testing for the linear independence of a set of solutions of second order differential equations. Let f_i represent a set of functions then the criterion for linear *dependence* is the existence of a relation of the form $\Sigma k_i f_i = 0$, wherein not all of the coefficients k_i are zero. On the other hand, if the only solution of the above equation is $k_i = 0$ for all i, the set of functions f_i is said to be linearly independent. Assuming that the functions f_i are differentiable then, differentiating the above equation repeatedly we get a set of equations.

$$\Sigma k_i f_i' = 0 \qquad \Sigma k_i f_i'' = 0, \text{ etc.}$$

Thus, we have set of homogeneous linear equations wherein

k_i are unknowns. Now there is a solution $k_i = 0$, only if the determinant of the coefficients of the k_i's vanishes, that is

$$\begin{vmatrix} f_1 & f_2 & \cdots\cdots & f_n \\ f_1' & f_2' & \cdots\cdots & f_n' \\ f_1^{(n-1)} & f_2^{(n-1)} & \cdots\cdots & f_n^{(n-1)} \end{vmatrix} = 0$$

This determinant is known as *wronskian.*

X

X -, Y-, Z- Axes. Syn. *Coordinate Axes.* Refers to the three axes in the three dimensions cartesian coordinate geometry. In the diagram the line X'OX, Y'OY, Z'OZ through the origin of coordinates represent the X-, Y-, Z- axes. Their equations are,

$$y = 0, z = 0; \ z = 0, x = 0; x = 0, y = 0.$$

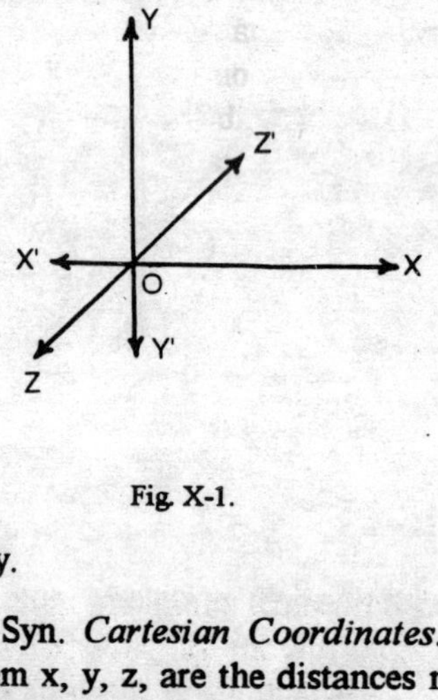

Fig. X-1.

respectively.

x, y, z Coordinates. Syn. *Cartesian Coordinates.* In the cartesian coordinates system x, y, z, are the distances measured upto the coordinate planes from the given in space.

x, y, z, Intercepts. In the system of certesian coordinates refers to the distances from the origin to the points of intersection of the curve on the surface with x-axis, y—axis, z – axis respectively.

xy, zy, zx- Planes. Syn. *Coordinate Planes.* In the cartesian coordinate system refers to the planes which contain the x-axis and y-axis and z – axis, and z – axis and x – axis, are the xy, yz and zx planes respectively. Their equations are given by z = 0, x = 0, y = 0 respectively.

Y

Yard. The british Imperial and U. S. standard unit of length. It is divided into 3 feet or 36 inches. It is also one thousand seven hundred—and sixtieth part of a statute *mile*. This has now becomes obsolete in science and engineering. One yard = .9144 metres.

Yates Correction for Continuity. The estimate of χ^2 for a 2 × 2 table needs a correction for small frequencies. The following formula for χ^2 contains a correction that results in reasonably close approximation to a χ^2 distribution when the expected number of cases in each cell of 2 × 2 table is small :

$$\text{Cor. } \chi^2 = \sum_{i=1}^{4} \frac{(|x_1 - m_1| - \frac{1}{2})^2}{m_1}$$

where x_1 is the observed frequency and m_1 is the expected frequency in the ith cell.

Yaw, Yaw Angle. In exterior ballistics, refers to the angle between the direction of the axis of a shell and the direction of its velocity vector.

Year. Refers to the period of revolution of the earth around the sun. The *tropical* or the *Solar* year is the time that passes between the two successive appearances at the first point of Aries. It equals 365.2422 mean solar days.

The Calendar year, is considered equivalent to 365.2422 mean solar days.

The Sidereal year is considered to be 365.2564 mean solar days. A lunar year, is equal to 12 luman months and is considered to be 365.3671 mean solar days. Three consecutive calendar years consist of 365 days and the fourth, called a *leap* year, has 366 days. A year divisible by 4 is the leap year.

Yield. Refers to the income produced by a stock or share expressed as a percentage of its market price. For example, if a share having a nominal value of 50p pays a 12% dividend it will pay 6p per share. If the market price is 80p, the yield would be $(6\times100)/80 = 7.5\%$. In the case of stocks that pay fixed interest, the yield to a large extent determines the market price. The buyers of stock will expect a yield that is comparable to the current interest rates; if these are rising the yield of fixed-interest stocks must also rise. This can only happen if the market price falls. Thus rising interest rates depress the market price of stocks, and vice versa.

Z

Zenith. Refers to a point on the celestial sphere (or in the sky) that is directly over the head of the observer. The highest point.

Zenith-Nadir Line. Refers to the line joining the *Zenith* to the *Nadir*.

Zero. In arithmetic, the *identity element* of addition i.e., the number 0 for which $x + 0 = x$ and $0 + x = x$ for all number x . Zero is also the cardinal number of the empty set.

Division **by Zero**. See Division.

Division **of Zero**. The quotient of zero and any nonzero number is zero; $0/k = 0$ for all k not since $0 = k \times 0$.

Divisor of **Zero**. See. Domain -integral domain.

Factorial **Zero**. Defined as equal to 1.

Multiplication **by Zero.** The product of zero and any other number is zero, i.e., $0 \times k = k \times 0 = 0$ for all k.

Zero of a Function. A value of the argument for which the function is zero. A **real zero** is a number for which the function is zero. If the functioin f has only real number values, for real number values of x (e.g., if f is a **polynomial** with real numbers as coefficients), then the real zeros of f are the values of x for which the curve $y = f(x)$ meets the x–axis. See Root —root of an equation. If z_0 is a zero (also known as **zero point**) of an **analytic function** f of a complex variable, then there is a positive integer k such that $f(z) \equiv (z - z_0)^k \phi(z)$, where ϕ is analytic at z_0 and $\phi(z_0) \neq 0$. The integer k is the **order** of the zero.

Zero-Sum Game. *See* Game.

Zero Vector. Refers to a vector of zero length; a vector all of whose components are zero. For vectors of type $V = ai + bj + ck$, the zero vector is the vector O = 0i + 0j + 0k. The zero vector is the zero element for vector addition, meaning that V + O = O + V = V for all V.

Zeta. Refers to the sixth letter of the Greek alphabet, written capital, Z.

Riemann **Zeta Function**. The zeta function ζ of complex numbers z = x + iy is defined for x > 1 by series

$$\zeta(z) = \sum_{n=1}^{\infty} n^{-z} = \sum_{n=1}^{\infty} e^{-z\log n}$$

where log n is real. The function can be defined by analytic continuation for all finite z. It is a meromorphic function having a simple pole at z = 1.

Zodiac. Refers to zone of the celestial sphere wherein the sun, moon and planets trace their paths. It is divided into 12 sections according to the apparent position of the sun in the 12 months of the year. These sections are designated by the sign of the Zodic: Aries (Rain), Tarus (Bull), Gemini (Twins), Cancer (Crab), Leo (Lion), Virgo (Virgin), Libra (Scales), Scorpio, (Scorpion), Sagitarius (Archer), Capricornus (Sea goat), Aquarius (Water-bearer), Pisces (Fishes).

Zodiac. Refers to the portion of a sphere that is bound by the two intersections of two parallel planes with the sphere. One of the planes may be a tangent planes, so that one of the circular intersections is a point and the zone is said to be a **zone of one base**. A **base** of a zone is an intersection having the sphere of one of the planes forming the zone. The altitude of a zone is the perpendicular distance between the planes cutting the zone out of the sphere. The area of a zone is equal to the product of its altitude and the perimeter of a great circle of the sphere, i.e., $2\pi rh$, where r is the radius of the sphere and h the altitude of the zone.

Zone of a Surface of Revolution. Refers to the portion of the surface contained between two planes normal to the axis of revolution.

Zorn's Lemma. The maximal principal. If T is *partially ordered* and each *linearly ordered* subset has an upper bound in T, then T contains at least one *maximal element* (an element x such that there is no y of T with $x < y$). Other alternative forms of this principle are : (1) (**Kuratowski's lemma**). Each simply ordered subset of a partially ordered set is contained in a maximal linearly ordered subset. (2) If a collection A of sets has the property that for each *nest* in A there is a member of A that contains each member of the nest, then there is a maximal member of A. (3) (**Hausdorff maximal principle**). If A is a collection of sets and N is a nest in A, then there is a nest N* that contains N and is not contained in any larger nest. (4) (**Tukey's lemma**). A collection of sets which is of *finite character* has a *maximal member*. (5) Any set can be well-ordered (see Ordered — ordered set). (6) The axiom of choice (see Choice). If the finite axiom of choice is assumed, all of the above principles are logically equivalent.